ENVIRONMENTAL

ADVOCACY

Concepts, Issues
and Dilemmas

Bunyan Bryant

Environmental Advocacy: Concepts, Issues, and Dilemmas
By Bunyan Bryant

Copyright 1990 by Bunyan Bryant

Published by **Caddo Gap Press**
    317 South Division Street, Suite 2
    Ann Arbor, Michigan 48104

Price - $9.95

ISBN 0-9625945-1-2
Library of Congress Catalog Card Number 90-082936

Caddo Gap Press

CONTENTS

CHAPTER 1

THE HISTORICAL context of environmental advocacy

Environmental Advocacy (EA) is a critique of both present conditions and history. It grows out of our understanding of historical forces that have played a significant role in shaping the world. Understanding history provides us with a context for understanding present conditions and how we relate to them. We start with the post-Civil War era, when the invention of the internal combustion engine and the allocation of millions of acres by the United States government to railroad companies increased the pace of industrialization, opening new markets and access to raw materials. Oligopolies in transportation, oil, iron, utilities, and banks were expanding and consolidating their power (Ash, 1972). As industry gained power over our natural resources and labor, the government often failed to enforce anti-trust and commerce laws; a kind of "Social Darwinism" prevailed, providing justification not only for exploiting workers, but also for treating them as inferior beings (Ash, 1972). This post-Civil War era was a time of cutthroat competition and militant labor organizing.

EARLY HISTORY

Incredible ecological destruction was wrought by the captains of industry as they exploited the nation's resources for private gain. In the early stages of the conservation movement, both the preservationists and the utilitarians articulated many of the same ideals and needs for nineteenth-century America, and worked together to preserve forests and parks from unenlightened exploitation and corruption (Petulla, 1977). To prevent wanton destruction of our nation's heritage, broad-based support manifested itself in conservation organizations--such as the Sierra Club, whose first president was John Muir; the Boone and Crockett Club, formed by Theodore Roosevelt; some lesser known hunting and fishing clubs; the Audubon Society; and many others (Humphrey and Buttel, 1982). When Gifford Pinchot became chief U.S. forester, a philosophical and ideological estrangement arose between Muir's ideals of forest preservation and Pinchot's scientific management approach--a split that became the focus of environmental conflict in that era (Petulla, 1977).

Muir, who was a preservationist, and the Audubon Society wanted the federal government, for historic, recreational, and scientific purposes, to reserve undeveloped habitats so that they could undergo ecological succession with minimal human interference. On the other hand, Pinchot, whose background was that of a forester trained in Europe, the Boone and Crockett Clubs, ranchers, and other consumptive wildlifers favored utilitarianism--i.e., reducing natural resource exploitation through scientific management. The conservation policies of Theodore Roosevelt attempted to accommodate both preservationists and utilitarians by exerting greater governmental control over both private enterprises and the public domain. Thus, the preservationists were able to pressure government into protecting recreational land and land of limited commercial value from the instruments of exploitation. Even so, the conflict between these two forces continues to this day (Humphrey and Buttel, 1982). For example, it is clear that universities and schools of forestry and natural resources are teaching scientific management for utilitarian purposes supported by industry; on the other hand, many nonprofit environmental organizations are more preservationist in character, critical of government policies and--in some cases--of universities.

landmark acts

Even though the struggle between these two factions has been intense, it has propelled the conservation movement forward. Some of the early landmarks of the conservation movement were the Yellowstone Act of 1872 which preserved an area of outstanding natural beauty and value for public enjoyment and scientific research, the establishment of the Adirondack Forests Preserve in New York state in 1885, and the passage of legislation to preserve the Yosemite area in 1890 and Mount Rainer in 1899. In addition to these groundbreaking events, the U.S. Congress passed the Forest Reserve Act in 1891, setting aside 1,250,000 acres of natural forests as a reserve, and the Forest Management Act of 1897, which allowed such uses of the Forest Reserve as mining, grazing, and lumbering with permits issued by the government. Today the Forest Service employs over 20,000 persons to manage over 180 million acres. These represent only a few of the accomplishments of the conservation movement--achieved despite the ongoing conflict between environmental philosophies.

The conservation movement, which included both preservationists and utilitarians, changed the political economy of the nation from one that was basically unregulated, in which industry had considerable flexibility in the exploitation and destruction of our forestlands, to one that sometimes required the government to undertake the management of our national heritage and protect it from wholesale exploitation by the captains of industry. By the turn of this century, the greatest, swiftest, most efficient, and appalling wave of forest destruction in human history had taken place. Of lasting importance was the change in the political economy, cementing a relationship between corporations and government whereby the government not only took charge of public lands for scientific management, but also undertook policies and activities such as land reclamation and irrigation. These activities were actually a subsidy to large businesses--at least more so than small ones. Forests and public lands were sold or leased to corporations at or below market value for purposes of scientific management.

strengths and weaknesses

It is important for environmental advocates to understand the strengths and weaknesses of the preservationist and conservationist movements. While the former is basically anti-development, the latter exploits publicly owned or subsidized resources for the differential benefit of elite groups. EA, on the other hand, both advocates a high degree of preservation and places more emphasis upon the prudent use of our collective heritage for the benefit of the many rather than the few. The political economy that was forged during the conservation movement helped create a relationship between government and industry that continues to this day. It recently became blatant when the Reagan Administration, through Secretary of the Interior James Watt, leased or sold off forest and mineral resources at firesale prices to corporations. "They handed over to private use clean air and water, forest and grasslands, coal and oil that belongs to us all. In the name of getting the government of our backs, they are giving away our natural heritage" (Friends of the Earth *et al.*, 1982:6).

ANTECEDENTS OF EA

In looking at the history of both the conservation and preservation traditions, it is important to understand that EA did not grow out of either of these traditions; EA did not take root until the 1950s, and did not flourish until the 1960s and 1970s. Important cultural, social, and political influences in those decades gave rise to associated movements.

The 1950s were marked by quietism, a conformity that characterized the post-World War II era. The little resistance that did take place during this time was blunted by economic growth or suppressed by McCarthyism. Even though apartheid and poverty reigned in the Deep South, and racial discrimination in the North, university students preparing themselves to ascend the corporate ladder were more preoccupied with the social life of fraternities and sororities than with the social cause of equality. But underneath this quietude, poets on the streets of New York and San Francisco began to critique middle-class values. Among the most famous of these writers were Allen Ginsberg and Jack Kerouac, two forerunners of the "New Left" (an extremely anti-establishment movement), who were also among the first to confront the contemporary values of American society. Through their poetry and novels, they encouraged a whole generation of young Americans to question middle-class values. The poets of the "Beat Generation" were not the only influence on youth, however. Rock and roll, a rhythmic new musical form that helped people release their pent-up emotions and lowered their inhibitions, also had its effect.

civil rights

Further social influences on contemporary attitudes sprang from surprising sources. In the winter of 1955, Rosa Parks, a black woman in Montgomery, Alabama, refused to give up her seat and move to the back of a racially segregated bus. This single act triggered one of the greatest movements in the history of this country. Over night, a little known minister named Martin Luther King became the backbone of a new movement and was thrust into national prominence. As the 1960s began, black and white students left the campuses to join the civil rights struggle in the South; they participated in nonviolent demonstrations, civil disobedience, and voter registration, and helped organize political parties such as the Black Panther Party of Lowndes County Alabama, and the Freedom Democratic Party led by Fannie Lou Hamer in Mississippi. While the Student Nonviolent Coordinating Committee (SNCC) concentrated most of its efforts in the South, the fastest growing organization of that time, Students for a Democratic Society (SDS), began to focus its efforts on organizing white communities in the North around economic issues and empowerment. Social experimentation became common, and though some experiments in communal living and "returning to the land" were successful, at least in the short run, the cry of "do your own thing" and oppressive practices in the name of change caused many such experiments to fail. Nevertheless, these experiences challenged student values and lifestyles in a profound way. When students returned to the college campuses with firsthand knowledge of the social injustices and inequities of our system, they proceeded to challenge university courses, rules, and any suppression of free speech.

the politics of the 1960s

The politics of the 1960s revolved around a war in Vietnam and a war at home. The Vietnam conflict brought thousands of people, of all racial and social backgrounds, into the nation's streets. This movement eventually slowed down the Pentagon's efforts to further escalate the war and discouraged President Lyndon Johnson from seeking another term in office. For a brief time, inner cities across the nation erupted in rage and violence. The city of Chicago exploded during the 1968 Democratic National Convention, as protesters against the war in Vietnam were beaten and arrested by police. While this domestic war raged, the conflict in Vietnam continued. The casualties and mortality rates were not limited to those inflicted directly in combat; the long-term health of veterans was frequently affected by the herbicides 2,4,5-T and 2,4-D and Agent Orange, used to defoliate the Vietnamese countryside for military purposes.

Further social and political influences developed in the 1960s, as consciousness-raising groups initiated a substantial re-evaluation of women's roles. A number of women authors began to question patriarchy, women's economic status, and the restricted and repressed social roles women had been relegated to play. Such forces as the student movement, movements for civil rights, peace, and equality for women, the counter-culture movement, and--later--the environmental and appropriate-technologies movements all questioned the values, assumptions, and legitimate power of the establishment. Activists began to see connections linking military domination, patriarchy, racism, and the domination of nature. This was the "New Left," which budded in the "Beat Generation" and flourished as a loosely coherent mosaic of multiple movements. It is from this tradition that EA springs.

groundbreaking books

Other events in the 1960s also rekindled interest in the values of the conservation movement which became known as the environmental movement. Rachel Carson's best seller, *Silent Spring*, carried an emotional message of the invisible dangers of post-World War II pesticides and herbicides. Partly in response to this renewed interest, intellectual efforts for environmental concerns expanded. Three intellectuals especially important to EA surfaced about ten years after publication of *Silent Spring*. Paul Ehrlich wrote a best seller entitled *The Population Bomb*, in which--in the Malthusian tradition--he theorized that world population growth was the cause of pollution and other contemporary social problems. Barry Commoner, who has written several books on the environmental revolution, criticized Ehrlich's thesis regarding the population explosion and resultant problems, giving evidence instead that both pollution and energy shortages have multiple causes. Garret Hardin's lifeboat metaphor, which hypothesizes an overcrowded lifeboat which forces ethical decisions to determine who lives and who dies, has been posed as a method for determining which developing countries should be saved from starvation. These authors have presented some interesting analyses of environmental problems and have triggered renewed debate and activism, allowing for enhancement of EA through both critique and employment of their conceptual thinking. EA finds more comfort in the works of Commoner than those of Ehrlich and Hardin, but even though EA differs with the latter two, it recognizes them for providing a platform for debate and for increasing consciousness of the issues.

The expanded debate initiated by these authors continued into the 1970s. *The Limits to Growth* (Meadows *et al.*, 1972) predicted ecological collapse if present trends continue in the growth of population, industry, and resource use, while *Mankind at the Turning Point* (Mesarovic and Pestel, 1974) and *RIO: Reshaping the International Order* (Tinbergen, 1976) both rejected the notion that the world system could be portrayed in terms of world level or averages. Mesarovic and Pestel state that the world should be viewed as a system of interacting regions, rather than as a unit, and that some regions will reach ecological collapse before other regions. The collapse of regions can be attenuated if the gaps between the "First World" and the "Third World" are closed by more "First World" investment aid to underdeveloped regions. The *RIO* study has a more positive attitude towards growth reflected in its calls for the world to reserve the use of untapped resources, to establish standards for the flow of resources, to conserve energy resources for the benefit of the

"Third World," and to implement international planning. Both of these studies were much more optimistic and included a broader range of policy choices than *The Limits to Growth*. The debate carried on in these bestsellers has made us more aware of the potential of world ecological collapse (Humphrey and Buttel, 1982).

Around this same time, an intellectual development began among a number of sociologists who had been observing the environmental movement and documenting how our social, political, and economic institutions affect environmental quality. They asserted that rapid depletion of our natural resources and the degradation of the environment would undermine the institutional structure upon which society is built. Our institutions cannot survive without being fueled by biological capital or critical resources such as energy from fossil fuels. Although most were trained as traditional sociologists, these thinkers began to challenge the anthropocentric point of view of Western sociology that views humans as distinct and separate from nonhumans. They challenged the assumptions of seventeenth-century Eurocentric thinking that practiced exploitation and domination over nature. To understand the conceptual aspects of social change and the environmental movement, they forced themselves to move their analysis from a "human exemptionalism" paradigm (this term is now used instead of the previous "human excep-tionalism") to a new environmentalist paradigm--one that views human life as on a continuum with nonhuman life, subject to the same laws of nature.

recent legislation

Much of the early environmental movement had its origins in mainstream efforts to conserve natural resources for the long-term use of elite groups. The extension of this struggle has resurfaced more intensely within the last three decades. By the middle 1960s, lakes and streams were literally dying and the lives

of people who lived near and far from them were threatened by
the dumping of sewage and industrial waste. This prompted
passage of a number of federal acts, such as the Federal Water
Pollution Act of 1972, the Clean Water Act of 1977, the Solid
Waste Disposal Act of 1976, the Toxic Substance Control Act of
1976, the Reclamation Act of 1977, and the Superfund in 1980.
Despite passage of this legislation to eliminate dumping of
hazardous and toxic waste, one still find thousands of sites across
the country where dumping takes place, many of which are
located in neighborhoods where a high percentage of minorities
live. The protests in Warren County, North Carolina, to prevent
the dumping of PCBs in a dumpsite located in a predominantly
minority neighborhood resembled the civil rights struggles of the
1960s (Alderman, 1982). More such protests are likely as these
poisons begin to affect people's health.

A WORLDWIDE CONTEXT

In the early 1970s, the energy crisis strained Americans'
pocket books and jarred the national consciousness. Suddenly
everyone was vulnerable to the oil-producing nations. Project
Independence and the Energy Mobilization Board were instituted
to support and develop energy resources in the western U.S.
Even though members of the Organization of Petroleum
Exporting Countries (OPEC) raised the price of oil and
nationalized oil resources and the assets of international oil
companies, little protest came from the companies, because they
still owned the shipping and refineries and had ultimate control;
the oil-producing countries were still dependent upon corporate
technology. Increased prices made coal in the western United
States competitive with oil, and energy companies began to move
west to exploit the nation's coal reserves at relatively low costs,
putting economic pressures upon ranchers and Native Americans
to lease or sell their land. Furthermore, in response to the energy

crisis, utility companies began to build more nuclear power plants with government subsidies. This was to be the wave of the future; good clean energy from nuclear power plants. But a storm of protest against nuclear power plants began to gather. The most famous protest took place at Seabrook, New Hampshire, where, in May of 1977, some 1,400 nonviolent demonstrators were arrested while trying to occupy a nuclear power construction site.

President Jimmy Carter referred to the urgency of the energy crisis as the moral equivalent of war, helping more people become aware of and committed to ecological values and a new way of thinking and living here on planet earth. The nation could no longer afford to use energy resources at exponential rates, because energy shortages caused by such use would undermine economic institutions; America could no longer afford to foul the global nest and expect to survive. The nation was asked to dial down its thermostats and to engage in energy conservation. But the moral equivalent of war was short-lived; the discovery of new sources of oil forced down OPEC prices and people returned to business as usual.

scholarly views

During this time, three dominant scholars offered or pointed the way toward solutions to the environmental crisis. In his book *Small Is Beautiful*, Schumacher (1973:55) described Buddhist economics in which the Buddhist "sees the essence of civilization not in a multiplication of wants but in the purification of human character," so that work can not only supply personal needs and luxuries but also provide the conditions for human dignity, freedom, and liberation. Lovins (1976) posited that solar, geothermal, hydroelectric, wind, cogeneration, biomass, and other underutilized forms of energy, combined with energy conservation, are alternatives to multimillion-dollar nuclear power plants for the generation of electricity. Rifkin (1981) warned about consequences of the exponential use of energy, of how energy moves from a useable to an unusable state, and of how the unavailability of energy could undermine social institutions. These perspectives reflected the counter-cultural and environmental values of the 1960s, and combined with an appropriate-technology movement offered potential solutions to many of society's problems by advocating the use of small-scale technology that is benign to the environment. This appropriate-technology movement has not yet entered mainstream debate; it has not taken into consideration the issues of social justice and economic equity in a meaningful way.

political directions

The environmental movement has, however, entered American politics. In 1980, Barry Commoner of the Citizens Party ran for President of the United States. This was probably the first time that a presidential candidate was both a socialist and a biologist. Although his campaign and the Citizens Party championed "green" issues, they were not able to build a broad-based coalition. Then, in 1984, the long political trek of Jesse Jackson and the Rainbow Coalition began. The civil rights movement entered a new phase by interjecting itself into electoral politics. By combining a new leftist agenda with religious metaphors and, to a somewhat lesser extent, green politics, Jackson was able to appeal to a broad-based constituency. He was able to challenge the Democratic Party in a significant manner, and to raise the conscience of a large part of the nation; he kept hope alive for millions of Americans, regardless of their race, color, or creed. He encouraged minorities and progressive whites to run for political office at all levels of government. The long-range impact of his candidacy has yet to be measured.

European developments

Environmental concerns have had an impact beyond the United States. The "Green Movement" in West Germany, a loosely-held-together coalition of peaceniks, environmentalists, liberal, conservative, and left-wing activists, has grown out of deeply held concerns about the potential destruction of Europe by nuclear weapons. The cold war military alliances that allowed the United States to deploy missiles in western Europe and the Soviets to do so in eastern Europe were signs to the Greens that the superpowers were willing to risk nuclear war on European soil. They realized that by agreeing to carry out U.S. foreign policy of containment, West Germany was collaborating in its own potential nuclear destruction, and possibly the destruction of the rest of the world. The movement gained international attention, turning out millions of people in resistance to the U.S. Pershing II and Cruise Missiles, as well as by electing their members to the West German Parliament. It seems at first glance that the Green Movement was indigenous only to West Germany; closer observation, however, clearly indicates that the genesis of this movement was in the United States. Petra Kelly, a key activist in the West German Green Movement, went to school in the United States and was much influenced by Martin Luther King, Robert Kennedy, and Hubert Humphrey. She and others in West Germany were also influenced by the civil rights and environmental movements; by the principles of ecology, grassroots democracy, and nonviolence; and by the wisdom of Native Americans and Henry David Thoreau. Although the Greens have drawn on other antecedents as well, their impressive achievements have grown largely from American seeds (Capra and Spretnak, 1984).

focusing EA

In the summer of 1987, a meeting of the Greens took place

in Amherst, Massachusetts. People came from all over the United States and other parts of the world, and the West German influence was much in evidence. Present were people with a variety of different political and ecological approaches. One of these points of view is "deep ecology." Arne Naess, a Norwegian ecophilosopher cited by Miller (1988:593), pointed out two perspectives to environmental problems in 1973--he defined "shallow ecology" and "deep ecology." The former corresponds roughly to the Spaceship Earth world view, and the latter corresponds roughly to the sustainability world. Deep ecology has had a profound effect upon the Green Movement. Emphasizing biocentrism, the principle of ecosystem-centeredness, deep ecology stresses that nonhuman life has a right to live; one does not have the right to destroy other species without sufficient reason (Miller, 1988:593). Of considerable importance in the Green Movement is this spiritual concept of our connectedness with all life forms, and the potential of discovering truth and meaning through self-reflection.

Other points of view represented at the Amherst conference emphasized being more aware of our sexual being, our bodies, and the freedom of sexual preferences within the context of a supportive environment. Ecofeminism has emerged as a critique of patriarchy, which embodies the domination and manipulation of nature with women serving a male-controlled hierarchical system (Sale, 1987). In the political realm, some believe that Greens should not become a political party and involve themselves in electoral politics, and others feel that they should follow the model of the West Germans who have been successful in electing people to parliament. Still others believe that the movement should dismantle present geopolitical boundaries that are artificial, and replace them with natural boundaries that follow the lines of sustaining ecosystems. All of these perspectives make unique contributions to EA. The Green Movement and party, Rainbow Coalitions, and environmental advocates are all necessary to chart a new course so that environmental disasters such as acid rain, global warming, depletion of the ozone layer, and industrial accidents like Bhopal, can no longer happen.

Environmental advocates must take time to learn from history, from the strengths and weaknesses of the analyses of professional and lay writers reporting on past events. Even if one disagrees with certain analyses of history, one must understand why there is a disagreement and how that disagreement can strengthen one's own analysis and action for a fairer and just world. If one fails to learn lessons from the past one will surely repeat the mistakes of history in the future. The world is growing more and more dangerous each day; it cannot afford too many more mistakes. If they occur, they will unquestionably be costly.

CHAPTER 2

A DEFINITION of environmental advocacy: critique, skills, and strategy

In 1972, the School of Natural Resources at the University of Michigan launched an innovative program called "Environmental Advocacy." In the years since, the program has taken on aspects of a unique philosophy, as well as offering academic courses to prepare students for work in the environmental field. Although EA is a critique of society, assessing governmental policies, legislation, sponsored research, military programs, and the by-products of production, it also evaluates the ways in which excessive energy is harnessed to build "things" or symbols. While this critique is consistently changing and evolving, it is not enough by itself. Skills and viable strategies are needed to move beyond critique to the empowerment of organizations, and the people within them, thus enabling them to make meaningful demands on corporate and government decision-makers. Both skills and strategies must be aimed at empowering people to engage in widespread efforts to alter fundamental relationships, thus bettering their lives and protecting themselves from deteriorating environments and detestable social conditions.

Over the years, a definition of EA at the University of Michigan School of Natural Resources has emerged that helps to explain the behavior orientation of graduates. This definition, which includes a critique, skills, and strategies, is by no means conclusive. It serves as an important guideline for graduates of the program and others interested in social transformation.

EA is a critique of political economy and culture. Political economy may be described as the engine that drives society; it is the interaction between powerful economic and political institutions that fosters asymmetrical relationships among groups in society; it is formally and informally a political contract to use parliamentary and administrative government to protect, support, and enhance the economic interests of some groups more than others; it is a way that elite groups, through their institutionally-based economic and political power, have chosen to distribute societal rewards and services.

GOVERNMENT
AND CORPORATE CRITIQUE

EA critiques government policies that benefit certain sectors of society more than others. Government often goes to extremes to change legislation or social and environmental policies so that powerful interest groups may exploit publicly-owned resources for private gain. Through legislative and policy changes, government often gives the private sector less restricted use of air and water. For example, in 1982, the United States government changed the clean air rules, allowing coal-fired generating plants to spew tons of sulfur dioxide into the air, where it is transformed into acid rain that destroys forests, lakes, and streams. Around the same time, the government adopted rules to allow less control over companies dumping toxic wastes into landfills, and federal agencies began making fewer safety and environmental inspections, thus prosecuting fewer violators.

The federal government is not only relaxing rules and regulations for polluters, but in 1982 it gave control of more than a three-year supply of uncut timber on public lands to lumber companies and leased 16-and-one-half billion tons of coal to private industry, enough to last two centuries at the present rate of production. At firesale prices, they subsidized the production of oil, coal, timber, and synthetic fuels--all for private gain (Friends of the Earth *et al.*, 1982).

Corporate bureaucrats decry the morass of nuclear power plant regulations, claiming that extensive licensing delays cause economic insecurity and fail to enhance public safety. Evidence shows, however, that the government not only removes many barriers to the construction of nuclear power plants, but continues to subsidize this industry with billions of taxpayers' dollars. In short, the government promotes and subsidizes scientific research and the reprocessing and storage of nuclear fuels, and in some instances relaxes nuclear power regulations for private gain. This only increases the possibilities of the proliferation of the world nuclear arms race. It is expected that by the year 1990 more than 40 countries will have nuclear capability--generating 15,010 Kg. of plutonium, enough to make 3,000 nuclear bombs (Gyorgy and Friends, 1979).

fiscal priorities

Huge amounts of sponsored research dollars have entangled universities in a complex web of control, threatening the autonomy of higher education by outside forces or powerful interest groups that will determine the character of research to be performed. Research monies are made available for production-oriented rather than impact-oriented scientists. Foundations, corporations, and government pay professors to develop new technologies for a growth-oriented society; they spend comparably little money to study the societal impact of such technology. Both government and corporations avoid this latter kind of research for fear it may lead to more environmental restrictions and possible production losses. Hence, meager amounts of money are provided to scientists for developing appropriate technology, technology that is labor intensive and environmentally benign.

In the early 1980s, massive military spending--some of which came from shifting money from social to military programs--basically subsidized industry. Billions of dollars came into corporate hands for developing weapons, many of which were obsolete by the time of their completion. Cost overruns that give corporations billions of dollars of taxpayers' money amounted to welfare for the prosperous. Military spending--dollar for dollar--creates fewer jobs than equivalent spending on civilian goods and services (Rankin, 1981). Missiles in silos gather dust; mass transit systems, on the other hand, provide services and stimulate the economy. In some instances, military spending involves community planning, which often determines local boom and bust cycles in communities by allocating or withdrawing billions of dollars in military contracts.

The energy-oriented complex of huge oil companies, uranium mining firms, utilities, financial institutions, defense contractors, large construction companies, top nuclear equipment manufacturers, big architecture and engineering firms with their respective subcontractors, and related lobbying groups remains today a major concern because of the tremendous power it collectively wields. This is even further complicated by the fact that oil companies own shares of uranium and coal resources through interlocking boards of directors. For example, Exxon Nuclear--a subsidiary of the petroleum firm--produces nuclear fuels. Smaller companies like Continental Oil and Kerr-McGee Corporation hold valuable uranium acreage, and the latter has an interlocking directorate with General Electric Corporation (Berger, 1977).

changing relationships

Because the penetration of capital-intensive technologies into our lives usurps jobs and causes high unemployment and worker dislocation, the old social contract between management and labor has been weakened. This is manifested not only in the loss of jobs and worker dislocation, but in the threats of corporations to move from state to state or even to distant foreign shores in search of cheap and non-unionized labor. Government policies of tightening the money supply in the early 1980s triggered the worst recession since the Great Depression of 1929, throwing thousands out of work and forcing labor to make wage concessions due to an oversupply of workers. The downward pressure on wages, along with high unemployment, enabled corporate power to discipline labor by forcing or threatening wage concessions and/or job losses. Organized labor is losing membership; it cannot compete with the internationalization of capital.

Public funds for transportation are also a subsidy to industry. The automobile companies, for instance, need thousands of miles of improved roads for driver convenience and potential car sales. At the turn of the century, the automobile industry undermined a rather efficient transportation system by purchasing and phasing out mass transit systems, making people auto-dependent (Commoner, 1976). Today, only a small percentage of transportation funds are allocated to mass transit; most are earmarked for building new roads or repairing those already in existence.

Welfare, a basic floor of protection against abject poverty, subsidizes industry by providing income maintenance programs for unemployed workers until market forces improve. Welfare maintains a ready army of workers for the benefit of local industry. Similarly, some of the most intense protests against federal cutbacks in welfare programs come from local landlords, businesses, and bankers. They, too, are integral parts of the welfare state, as they profit from welfare-dependent clients who spend their monthly checks on rents, groceries, and loan payments. Welfare is not just for the poor, it is for both industry and the small business community.

exposing discrimination

EA is a critique of a political economy that differentially responds to people on the basis of their social class and racial background. Workers who live close to their work place are often vulnerable to pollution that damages health and property. For example, "the Appalachian coal miners who develop 'black lung,' the Southern textile workers who contract 'white lung,' and the California migrant workers who are exposed to high dosages of pesticides all work in debilitating environments" (England and Bluestone, 1973). Hence, as a group, these workers not only pay billions of dollars each year in medical bills and property improvements, but they often feel trapped in polluted and unsafe neighborhoods and work places, unable to move to safer and cleaner areas. Affirmative disaction and insidious institutionalized policies of government and private industry keep minorities locked in the debilitating environments of ghettos. These areas are characterized by factors such as: high population density and poor medical services; inadequate sanitation and garbage removal; poor street cleaning and rodent control; insufficient public transportation; disruption of neighborhoods for freeway construction; high noise levels; and electric rates higher than in the suburbs. Furthermore, toxic dumpsites are frequently located in neighborhoods that have a disproportionate percentage of blacks and working-class people. The lack of political will to respond to white-collar crime and crime against persons in any meaningful and fundamental way threatens all of our freedoms. United States farm policies that lead to family farm foreclosures and the concentration of land in the hands of the few also threaten the nation's way of life.

The careful construction of a complex web of governmental

and economic forces that determines what goods and services are produced and how they are distributed has not been without cost to health and the environments in which people live and work. This critique gives an idea of how the political economy is structured to support sustainable environments and people who live within them. Yet, political economy is only one part of Western industrialized culture. Thus, the critique must also examine culture's impact on society.

CULTURAL CRITIQUE

The anthropological work of White (1949) defines culture as society's ability to harness energy for "symboling," i.e., the ability to create mental representations or the exercise of mental abilities that are reflected in external reality. Examples of these mental representations are the design and construction of buildings, automobiles, furniture, stock and bonds, money, computers, clothes, robots, and, more generally, tools and technology--physical creations, both large and small, made by people. Human-made things are mental representations that convey information about culture.

harnessing energy

One important and distinguishing factor among cultures is their differential abilities to harness energy for growth and development. For example, in primitive societies the inability to harness energy beyond burning wood determines the simplicity of the social structure and culture. In such simple societies, considerably more land is required to sustain immediate and basic human needs. The inability to harness large amounts of energy ties primitive cultures directly to the land for survival.

In contrast, the ability to harness massive amounts of energy from coal created the conditions for the Industrial Revolution, enabling first England, and then other nations, to evolve into more complex societies. Harnessing extraordinary amounts of energy led to the creation of multiple forms of symboling, creating conditions and opportunities for the industrial nations of the West to exploit the lion's share of the world's natural resources.

The ability to harness energy also attracted people to smaller land spaces and changed their lives dramatically. Dense populations required new and different services, and the jobs created to fulfill these new requirements became once, or even several, times removed from the land. To show how such urbanization affects a society, let's take an example from U.S. history. In the 1920s, approximately 30.1 percent of the people in the United States farmed and made their living directly off the land, while by 1970 only 4.8 percent were engaged in such activities (Perelman, 1976). Now that the farm crisis of the 1980s has taken its toll, the number of people farming the land is probably less than 3 percent. The resulting overcrowded cities, traffic congestion, lack of adequate parking, multilevel buildings, and polluted environments characterize most industrialized cities.

the social contract

Culture plays an important part in one's life in other ways, too. It helps the species survive. With it comes a social contract--norms or values that make human behavior predictable. This contract sets forth guidelines for behavior to help sublimate or compensate for aggressive activities. While society harnesses more and more energy, making it convenient for population growth and crowded conditions in cities, the strain upon the social contract increases. This becomes evident in the breakdown of social control, which, in turn, requires increased dependency on the legal system to curtail violence and destructive behavior.

Another part of the American social contract is an emotional commitment to such high ideals as democracy, nationalism, freedom, and humanitarianism. Although democracy is embodied within the United States *Constitution*, people automatically assume principles of democracy without necessarily being reminded of democratic institutions. The social contract and humanitarian principles are often reflected in citizen response to emergencies and conditions of the poor and less fortunate. Democracy, freedom, and nationalism often motivate citizens to protect these principles and the country at all costs. Culture, in this way, is undoubtedly another major engine driving society. Perhaps, at one level, the ideals of culture are the glue holding society together, providing behavioral prescriptions for the survival of the species.

adaptive and maladaptive

While much of culture has been adaptive, allowing humans to symbol, to create technology, and to live and survive anywhere on earth regardless of climatic or geophysical conditions, a part of the culture has now become maladaptive in many different ways. (One could also say that the political and economic system is maladaptive because of its insatiable demand for growth and development, which transforms and destroys natural resources at an unprecedented rate.) Rainforests, which regulate global temperatures, rainfall, and provide oxygen and the habitat for 50 percent of the world's species, are cleared for raising cattle to provide beef for Western markets. The inefficiency of such land use becomes evident when it is realized that it takes 21 pounds of protein grain to produce one pound of protein beef for human consumption (Lappe, 1971), and that the same amount of grain would feed one person for a long time to come. Yet, much of the world starves to death or faces severe malnutrition while the more fortunate dine on steaks.

Inefficient land use is not the only maladaptive practice that characterizes agriculture. Cancer-causing chemicals are leached in the soil and streams. Organic and inorganic run-offs from feedlots and from fields treated with chemical fertilizers, herbicides, and pesticides increase eutrophication and lead to the destruction of aquatic life in lakes and streams. It is not enough for cancer causing chemicals to find their way into the food chain in this manner, but chances for ill-health are further increased by adding chemical-based additives directly to food. Furthermore, the kind of monocropping used in the U.S. has been responsible for the erosion of millions of tons of topsoil, creating long-term desertification and the potential for famine, not only in America, but in other countries as well.

Poisons, particulate matter, and chemicals released into the atmosphere alter world climates and create holes in the ozone layer, allowing ultraviolet light to penetrate and making people more susceptible to cancer. Combinations of various chemicals in the atmosphere provide the conditions for acid rain that destroy forest land, rivers, streams, and buildings. Leakage from toxic and hazardous dumpsites befoul deep underground aquifers, which then take hundreds of years to cleanse themselves.

More specific evidence of cultural maladaptivity is found in such further events as the nuclear accidents at Three Mile Island and Chernobyl, killing and injuring thousands of people and putting hundreds of thousands of acres out of food production; the problem of disposal of nuclear waste; the PCB incident in Michigan in the early 1970s, when fire retardant was mistakenly fed to cattle and entered the food chain where it can poison people for years to come; and the Berlin-Farrow dumpsite near Flint, Michigan, where leakage into the underground water table and fumes from the site affected the physical and psychological health of local residents. These are only a few incidents that mark the environmental landscape and threaten health and life. EA is a critique of such maladaptive culture, concerning itself with conditions such as those discussed here--conditions that can lead to "global cancer."

understanding energy

EA required that one become more aware of humankind's highly intensive energy use, and understand that without change society shall surely perish. There are, however, cultural mindsets that interfere with thinking about energy use and the environment. For example, the Newtonian worldview seeks to solve world problems by reducing them to linear equations, even though the complex web of biophysical environment does not lend itself to such simplistic linear notions. Furthermore, anthropocentric culture places humans at the center stage of the universe, reinforcing the notion that natural resources and nonhuman life exist for wanton exploitation. The growth ethic is at the basis of the ecological crisis; the exponential use of energy resources leads to entropy, i.e., a state in which energy changes from usable to unusable forms. Because energy is the foundation upon which society is built, its exponential use or its depletion threatens both social and economic institutions.

"DEEP ECOLOGY" CRITIQUE

EA is more than a critique of political economy and culture; it is also a critique of environmental and ecological movements and organizations, and thus of people that clothe themselves in an environmental ethic or use environmental concepts for their own purposes. One such movement is "deep ecology," a form of biocentrism that gives equality to all living things, from the tiniest microbes to the most complex living forms, regardless of the consequences. EA takes exception to those who take this position to the extreme. To give the AIDS virus the same status as other plant and animal species, claiming that it helps control world population, is the height of irresponsibility.

A few deep ecologists would further contend that the worst thing one could do is give aid to starving Ethiopians without letting nature take its course. Can anyone reasonably advocate "letting nature take its course" when Ethiopians and other peoples of Africa and of Latin America find themselves in conditions that are the result of centuries of colonialism and exploitation by the West? Nature had little opportunity to take its course in these countries when Western intervention, domination, and exploitation of natural resources ultimately led to relations of structural dependency and conditions of abject poverty. To let nature have its way, allowing people to starve, after hundreds of years of Western tinkering with nature is the height of conceit.

defining "Deep Ecology"

In order to continue with this critique of deep ecology, it is necessary to further define it. Naess (cited in Miller, 1988) and Devall and Sessions (1985) are the major proponents of this concept, which finds its roots in a number of traditions. One source of thought in deep ecology is Taoism, an Eastern philosophy that espouses a way of life based on compassion, respect, and love for all things--a compassion that arises from self-love of a larger self. Some tenets of Deep Ecology are bioregionalism, biological diversity, and biocentrism (as opposed to anthropocentrism). Deep ecologists maintain that to be detached from nature robs people of their unique spiritual/bio-logical personhood. No one can survive or be saved until all are saved, including grizzly bears, rain forests, ecosystems, mountains, rivers, and the tiniest microbes in the soil. They contend that if people harm the rest of nature, they harm themselves. There are no boundaries; everything is intricately related. No one has the right to destroy other living things without good reason. Although the supporters of Deep Ecology do not advocate going back to the stone age, they do urge reverence for the land, for primal people, and for communal societies, based on mutual aid and a bonding with nonhuman nature.

People who embrace Deep Ecology advocate an interconnectedness with all living things; they believe in a self that extends beyond ego, to include the world in which all live. They advocate returning to live among the animals as a way of completing the maturity cycle of humans, so that humans will feel less aggressive and biocentrism becomes the extreme. Yet, many of them would probably refuse to disclose their stock portfolios in economic institutions that are responsible for exploiting human and nonhuman resources. Nor would they abdicate their inheritance right. And though they advocate living among the animals, few are practicing what they say they believe. Few live in ghettos or among diverse human populations.

combatting dominance

EA is a critique of modern Malthusians who claim that famine, plagues, and war are good for population control. Such theorists say that before the world's population cascades out of control even more than it already has, one must find some way of controlling the population in less developed countries (LDCs), where populations will double in the next few years. Yet, it seems that a disproportionate amount of attention has been focused on LDCs, particularly in light of the fact that in Western Europe there are more people per square mile than in most LDCs, and those people probably use several times the resources per person than people in less developed countries. Although many have become experts in shifting the blame for world problems to people of color, claiming that they have too many children and will be a threat to world resources, they fail to see that the problems are rooted in world capitalist domination and exploitation by Western Europeans. Commoner (1975) in his article on "How Poverty Breeds Over Population" was able to show that as countries increase in affluence and education there is usually a drop in population. And Bookchin (1987:18) states that "if we provide people with decent lives, education, and a sense of creative meaning in life, and, above all, free women of their roles as mere bearers of children--and population growth begins to stabilize and populations rates even reverse their direction..." it will definitely make this a better world. Furthermore, Lappe (1977) states that while there is enough food to feed the world's population, many of the world's people still starve. Their starvation has to do more with who owns the land, what they do with it, and the lack of infrastructures for distribution than with uncontrolled population growth.

EA, like social ecology, "does not accept a biocentricity that denies or degrades the uniqueness of human beings, human subjectivity, rationality, aesthetic sensibility, and the ethical potentiality of this extraordinary species" (Bookchin, 1987). By the same token, EA, like social ecology, "rejects an anthropocentricity that confers on the privileged few the right to plunder the world of life, including women, the young, the poor, and the underprivileged" (Bookchin, 1987). EA is against dominance of nature and humans to a point where the few lord their wishes over the many and where the few do irreparable damage to spaceship earth.

THE COUNTER-CULTURE/ APPROPRIATE-TECHNOLOGY MOVEMENTS

Even though culture has served as an adaptive function during most of humankind's time here on earth, it has recently become maladaptive. The anthropocentric nature--the fascination with technology, growth, and world dominance--has taken humankind to the brink of nuclear disaster and to the despoliation of the environment and endangerment of health. EA feels that there can definitely be a better way for people in this world to relate to one another and to nonhuman nature; EA submits that the better parts of the adaptive culture can be cherished and embraced. It is up to all humans to humanize themselves and to be more respective of nature; it is up to all to change the materialistic ways that destroy the world's biological capital and to design a new life-style where basic needs are met and each person, regardless of race, color, creed, or sexual preference, can realize their highest human potential within an environmentally sustainable future.

a counter-culture movement

A counter-culture movement has emerged in the last two decades, criticizing basic values and assumptions underlying the conduct of one's personal life. It not only critiques the corporate community that hucksters products people neither need nor want; it also critiques the values inherent in certain symbols or high technology, as well as the exponential use of nonrenewable resources. It critiques the growth ethic, cogently arguing that the wanton destruction and misuse of natural resources undermine both social and economic institutions. And, perhaps most importantly, the movement presents alternative strategies and lifestyles, calling for greater integrity in personal and ecological lives.

Portions of the EA critique grow from writings of various counter-culture authors and environmental sociologists. These authors have written extensively on appropriate technology, energy and energy conservation, and the like. Schumacher wrote in *Small Is Beautiful* (1973) that appropriate decentralized technology is often more advantageous for society, in the long run, than large, energy-intensive, inappropriate technologies. Lovins (1976) questions the building and use of nuclear power plants, suggesting that solar and other alternative forms of energy are often more benign to the environment and draw less upon nonrenewable resources. Hayes (1976) states that half of all the energy used can be saved while still maintaining the present standard of living. Rifkin (1981) shows that the exponential use and long-term impact of the depletion of nonrenewable energy resources upon social institutions can be considerable. EA also seeks in this cultural critique to advance the understanding of culture itself and its influence upon both societal institutions and the bio-physical environment.

a vision of appropriateness

Based on the work of Schumacher, Lovins, Hayes, Rifkin, and others, a vision of what society should be begins to emerge. This vision is ever-changing and responding to the times, so an absolute vision is not appropriate here or elsewhere. What is being offered are parts of visions that will be integrated into a conceptual whole--a new vision that is just, meaningful, and supportive of an environmentally sustainable future.

EA has a vision of a society in which energy conservation will become standard practice rather than the exception, in which people will use appropriate technologies in decentralized, democratically controlled communities, in which a variety of cooperatives and worker-controlled industries will prevail, in which a new definition of work will allow people the choice of not working and still be supported by a modest income, in which community land trusts are created and used to release property from its speculative value, thus controlling the rate and direction of growth, and in which status is based on merit rather than inheritance. EA has a vision of a society in which solar energy is used to heat houses and electricity is produced by cogeneration and small hydroelectric plants where feasible. EA has a vision of society that is bioregional in character and more biocentric in nature. EA envisions a society where people demonstrate a planetary loyalty and a code of citizenship that allows for the deepest respect for all humans and nonhumans and a society where people can be loving and caring of one another, regardless of their race, color, creed, sexual preference, or national origin. EA dares to have a vision.

CHAPTER 3

ENVIRONMENTAL ADVOCACY and environmental mediation: differences and similarities

Environmental Advocacy (EA) is seldom embraced by those at the center of decision making, and, as a concept, it has been viewed as partisan and nonacademic. On the other hand, Environmental Mediation (EM) is gaining more legitimacy in academic and corporate circles. Because the process of EM is professed to be value-free and nonpartisan, it has been widely accepted by environmentalists, lawyers, and corporate executives, and has therefore managed to gain a position in the university as a legitimate and objective social science enterprise. While various universities have embraced EM in the pursuit of theory and praxis, the University of Michigan School of Natural Resources offers both mediation courses and EA as a field of study. The EA field of study is the only one of its kind in the nation. EA at the University of Michigan School of Natural Resources has survived as an academic program, but not without the skill and determination of its core faculty.

Because the courts are characterized by a backlog of environmental suits, tedious legal processes, delayed legal decisions, and great expense, EM becomes attractive to contending parties as an alternative to institutionalized legal means for solving environmental conflict. Because of the greater demand placed on the legal apparatus to solve social conflict, the very act of going to court as frequently as is done in this nation abdicates fundamental rights; it abridges certain freedoms, as more and more environmental activists rely upon decisions by judges and lawyers, many of whom are not elected officials. The legal system is not only expensive for users, but heavy users tend to put key decisions in the hands of officials once-removed from the people. Over the years, a complex dinosaur-like bureaucracy has emerged to handle lawsuits, many of which are questionable with respect to adjudication.

The previous chapter defined EA and illustrated it through various examples. This chapter will elaborate on that definition by explaining EA in juxtaposition to EM. Although EA and EM are both social change strategies, they have differences as well as similarities; they are different in their analysis of concrete conditions and the political context in which critical understanding is rooted. Yet, they may be similar in tactics.

THE POLITICAL ECONOMY
OF ENVIRONMENTAL MEDIATION

While EM operates within the political and economic context which is held together by consensus with respect to democracy, freedom, capitalism, growth, and productivity, the outcome solutions of environmental conflicts are usually predetermined by an overall consensus of a political democracy--not an economic democracy. Such solutions are predetermined by an economic system that is less than amicable to environmentalists, the working class, and the poor. It is the value-latent political economy that thwarts and frustrates any meaningful alternative paradigms; long-term solutions to the major underpinnings of social and environmental problems are not welcomed in this setting. Although environmental mediators claim value neutrality in solving environmental conflict, they fail to observe that they work within an overall system that is less than just and fair. Even though the idea of democracy is embraced along with all those nouns that are supposed to be a part of the American creed, issues of lineage and inheritance, of government catering to special interest groups, and of corporations profiteering outside the law make the society less than just. Obviously, Americans can say that they are better off than most people of the world. That may be true. But shouldn't one judge the country by how well it treats the aged and the young, women, the handicapped, the poor, minorities, and the biophysical environment relative to the social and economic potential within the context of a sustainable future.

Since the political economy has to do with the distribution of power and authority and the allocation of resources, power is often found to be the basis of corruption of elite groups, as evidenced by the amount and extent of white collar crime. For those located at the lower rungs of the political economy, their powerlessness is also a basis for corruption, as evidenced by the amount and extent of their alienation and despair, as well crimes against each other. Generation after generation of elite groups pass on their power, with its potential for corruption, to their heirs. And all too often, low-power groups pass on their condition of powerlessness and its potential for corruption to succeeding generations. Because power in this society is concentrated, elite groups are able to make decisions in their own best interests--often at the expense of low-power groups.

a local focus

By dealing with local issues such as the Grayrocks Dam case in Wyoming, the Houston River, and the Tennessee Eastman Company's debate over the limits to be set on the company's discharges (Pearson, 1987), EM seldom places the debate or mediation within the larger context of continued growth and exploitation of nonrenewable resources. Neither are the long-run effects or the byproducts of production on health, social concerns, or people's lives considered. For example, research has shown that toxic dumpsites are usually located in neighborhoods with high percentages of minorities and/or low-income areas (Bullard and Wright, 1986; Bullard, 1987; Commission for Racial Justice of the United Church of Christ, 1987). Yet such racial or structural factors seldom become part of EM's agenda.

Because EM assumes that the basic underpinnings of society are supported by the masses, mediating within this context is plausible; environmental mediators will never have to question the issues of fairness within the larger context, particularly if contending parties realize a certain degree of satisfaction. Fairness of outcome is already defined within the context of a political economy that benefits elite groups. The question is: Can there be fairness of a mediated agreement if the context in which it takes place is less than fair? Looking at fairness within the micro context often prevents the mediator from viewing the larger picture, with all its unsavory ramifications.

But isn't the most important thing that the parties involved feel that fairness has been rendered? Perhaps it is fair with respect to their subjective self-interests. Perhaps it is fair when contending parties feel that, from the struggle, they accomplished something of note. But culture mediates institutionally- pre-scribed expectations, or provides the lens by which one views the world and one's position within it. This, in turn, helps bring into sharper focus subjective self-interests. Even though mediated solutions may appeal to immediate or subjective self-interests, they may not appeal to long-term interests. If, for instance, one altered the political economy to be more just and human to both plant and animal species, then dispute settlements within this context would be fairer at both the micro and macro levels because it would have been done within an overall context that is more just and humane. It would have been done within a context where the objective long-term interest was seriously considered and dealt with.

broader EA view

EA, on the other hand, with its diagnosis and actions, is consistently conscious of the political economy and its mediating effects upon social institutions and environmental problems. Although society is held together by consensus decisions around such ideals as freedom, democracy, and civil rights, one knows in actuality that, even though such ideas exist in the hearts and minds of people, the discrepancy grows wider between those ideals and the rights and economic privileges of the masses. The nation is being propelled down the path of a two-tier society of "haves" and "have nots." Through environmental exploitation and government catering to powerful interest groups, it is clear that more and more of the wildlife and public lands will be privatized in the interests of elite groups and captains of industry.

As environmental advocates work to solve local or micro problems, EA places such problems in a larger context; it attempts to show the connection between the social and environmental problems that people experience in their neighborhoods and on their farms with national and international influences. Not all social and environmental problems can be laid at the doorsteps of current United States policy; Europe and Japan now also compete in and for world markets, including competition for American consumers. Yet, U.S. policies have caused a disproportionate amount of economic and social hardships for the working and lower classes and between the races. Lack of planning that allows industries to summarily close shops and move to distant ports for cheaper labor and resources leaves American communities in poverty and despair and destroys career aspirations. Automation and computer-driven machines are replacing workers at an alarming rate. Farm policies have caused foreclosure on the property of thousands of American farmers. Military spending and government-initiated recessions are providing benefits to corporate elites more than to workers; regaining control over inflation and interest rates comes at the expense of the working class and the poor.

seeking fundamental change

Without taking the political economy into consideration, it is most difficult to address issues of long-term equity or fairness. Without considering the long-term objective self-interest of the working class or the poor, EM as a problem-solving technique not only takes place in a void of understanding of those macro forces, but it seldom can be helpful to facilitate broad-based social change. Thus, EM is conservative in character, supporting only incremental change and protecting the status quo.

While EM empowers contending parties to solve site-specific problems, EA empowers them to engage in more fundamental change. As people struggle collectively to solve their problems, EA struggles to help enhance the consciousness of those who experience the negative effects of antagonists. EA uses struggles and victories, no matter how small, to help build and empower constituency groups to make meaningful demands upon those at the center of decision making. Although the task may be difficult, EA encourages not only immediate action and struggle, but it attempts where possible to lift that struggle to a level that confronts more fundamental societal issues. Thus, environmental advocates ally themselves with oppressed groups as opponents to well-established power and authority, seeking fundamental change of the status quo.

TRUST AND COMMUNICATION

Assuming that contending parties emotionally support the present political economy and assuming also that they are present at the bargaining table because they want to work out a solution for their best interests, mediation may well take on the characteristics of problem solving or collaboration. But in order to get to this stage, the mediator must engage contending parties in various trust-building activities. Such activities provide the context in which problem-solving activities take place. By building trust among conflicting groups, communication becomes less strained, stereotypes become less important, and clear solutions to problems become more possible.

Yet, it is the trusting building and congenial atmosphere created by mediators that could, in fact, disarm and coopt environmentalists or seduce them into compromising their ideals. Such trust building may cause environmentalists to lose their edge in championing both their subjective and objective self-interests, because of induced amicable relations with foes. And often trust building does not take into consideration the asymmetrical power relations between environmentalists and their opponents, who more often than not have access to numerous resources such as computers, data analysts, and/or a variety of consultants to empower them with knowledge in their interactions with environmentalists. Or mediators may try to reduce complex value-laden social and environmental issues to simplistic terms, thus disempowering environmentalists by making them look unreasonable.

communication skills

In EM's failure to question the role of the political economy and its effects upon micro mediation, there will undoubtedly be a heavy reliance upon developing communication skills, the assumption being that social and environmental problems can be easily solved if contending parties could clear up misunderstandings among themselves. Misunderstandings may well be a part of the social discourse of such conflicts, but to reduce social conflict to articulating misunderstandings represents a failure to deal with the underlying social issues. Contending parties may be--indeed, most likely are--in conflict not because of miscommunication, but because they acutely understand each others' position, and just plain disagree. It is as simple as that.

Confronting and challenging antagonists to change their behavior or their social or environmental policies does not leave much room for trust, because of their structural, economic, and political positions. Yet trust does play an important role in the relationship that environmental advocates have with their constituency groups, because those groups need to believe that environmental advocates will work to advance their cause or particular interests. To do otherwise would only hamper the effectiveness of EA work with constituency groups.

fundamental differences

While communication is important to environmental advocates in their work against antagonists, the emphasis is more often upon basic or fundamental differences than upon miscommunication. Conflict and communication become the curriculum for educating constituency groups about differences between themselves and antagonists, thereby understanding ways in which the struggle can be waged. Communication may serve very different purposes for EA than for EM. While trust and communication among contending groups may be important to EM, it is less so for EA. Because of the unequal distribution of wealth, income, resources, power, and authority, environmental advocates rely less upon trusting antagonists and more upon empowering constituency groups to make antagonists more accountable. Because antagonists are often power groups with great wealth and resources, and are instrumental in the destruction of the biophysical environment, trust--within the EA context--is seldom as high a priority as it is in EM.

Although many antagonists champion environmental protection, they continue to hold investment portfolios in companies that pollute the environment; they continue to gather wealth in the face of abject poverty; they resist high taxes and, in fact, advocate tax write-offs, tax breaks, and the like. Given their structural position and political and economic power, and given that Americans live in a highly competitive and growth-oriented society, the issue is not one of trust or miscommunication--rather, it is one of justice and redistribution.

If the issues were trust and communication, then the issue of redistribution of power and resources would have been settled a long time ago. People with power and wealth will protect their interests at all costs. The power elite seldom want to be accountable to the masses if they can help it. Democracy becomes a smoke screen for elitism.

SOCIAL-CLASS BACKGROUNDS
OF CHANGE AGENTS

Coming from the middle class, being schooled in some of the most prestigious universities, gaining legitimacy by various mediation and arbitration associations, environmental mediators claim to be value-free, neutral, and objective. To be effective mediators, they must embody a sense of fairness and objectivity--to do less would only tarnish the credibility of their value neutrality among contending parties. Yet, due to their social-class backgrounds, it is frequently impossible for them to be value-neutral and objective; mediators consciously or unconsciously lend their emotional support to procedures, processes, decisions, or opportunities that favor those that are close to their own structural position. Environmental mediators speak a language containing words that are seldom part of the vocabulary of low-power groups; indeed, often low-power groups feel estranged or even intimidated by language they do not understand. Thus, value neutrality and objectivity are myths that help legitimize EM. While environmental advocates may come from middle-class backgrounds, they are seldom trained in prestigious institutions, except for the University of Michigan. Most people become advocates out of strongly-held beliefs; they become advocates in an attempt to "right" certain "wrongs."

clarity of issues

Environmental advocates not only stake claim to an issue and bring resources to bear on it, but they are up-front with respect to who they are and where they want to go. There is no pretense to be value-free, neutral, and objective. They are clear on where they stand, and with that comes a certain honesty and integrity. Because a lot of environmental advocates are from middle-class backgrounds they are sensitive to language, discourse, and the "isms," particularly when working with low-power groups. EA not only attempts to deal with the "isms," but with content such as biology and political economy, within its program of courses.

While environmental mediators are more likely to use words and phrases such as conflict, collaboration, problem solving, trust in the process and each other, contending parties, consensus, mediation, and compromise, environmental advocates are more likely to use such words as antagonist, enemy, conflict, constituency building, and struggle. EM may also use data gathering by contending groups, either collectively or individually, as a way for each side to explore more acutely alternative possibilities of solving social or environmental problems. Engaging in the data-gathering process may uncover new issues and potential solutions, making predetermined solutions of contending parties obsolete. The emphasis is upon solving the site-specific problem.

empowerment instead

While environmental advocates use site-specific issues as a basis for social change, the emphasis however is not on site-specific solutions, but instead on empowerment or organization building for larger future struggles and outcomes. An issue that encourages the use of research to enhance the empowerment of an organization is important for long-term social change. Site-specific victories may not be important within themselves, but rather because of their potential to help build more powerful organizations. Such victories may not only provide a progressive critique of the political economy, but may help establish a better position from which to implement social change.

CONVERGENCE OF TACTICS

Even though EA and EM have fundamental differences regarding assumptions about society and intentional social change, they use many of the same skills and technology for their different ends. Mediators use communication strategies, group process, and various methods of problem solving, decision making, long range planning, data gathering, and diagnosis for settling conflicts. Advocates may not only use these skills for different ends, but because broad-based social change is so difficult, they also are required to develop a broader range of skills in order to be successful.

There are times, however, when advocacy for the empowerment of a constituency group reaches a point at which negotiation becomes the last resort for accomplishing or salvaging at least something from the struggle. It is not that EA is being coopted or selling out in such situations; rather, it is simply trying to get as much mileage from the effort as possible. Neither mediation nor negotiation is perceived as an end in itself, but rather as a way to enhance the power of constituency groups and to build organizations that can take on even larger issues in the future. When EA negotiates, it is to win small victories to enhance the collective, long-term power of the organization; it is to build organizations that question the fundamental assumptions of a system that is less than just; it is to use organizational power to push for fundamental changes for a more humane and equitable society.

Obviously, the task environmental advocates have taken upon themselves is more tedious and difficult than the task of environmental mediators. There are, however, some similarities in tactics as well as major differences in the understanding of issues, problems, and potential solutions. For environmental advocates, the search for fundamental change is an ongoing process. This will undoubtedly continue to be a high priority as EA struggles for justice, peace, and a sustainable future.

CHAPTER 4:

BUILDING ideology and organization for social and environmental change

In previous chapters, the definition of environmental advocacy and how it differs from environmental mediation has been discussed. In this chapter, more time is devoted to consideration of ideology, to how it helps environmental advocates explain themselves to each other and to the rest of the world, and to how it helps organize thoughts and provide a prescriptiveness for behavior.

Although many tend to think of ideology as dogma, EA takes issue with that; EA ideology, though grounded in action and reflection, is not carved in stone; it is subject to change with experience, action, and reflection over time. All students coming out of an EA program do not necessarily ascribe to the same ideology; but they do leave with many ideological similarities.

Ideology helps define who one is and what one stands for; it is this clarity that helps empower one to build organizations for social change. Being firmly grounded in a continuously evolving ideology helps one to become more resolved in initiating intentional social change. In fact, some individuals have been doing environmental advocacy work for years and will continue to do so. In this chapter, the discussion is of various aspects of ideology and how they relate to organizational building skills and professional leadership.

ORIGINS OF IDEOLOGY

Ideology is generally defined as "the manner of thinking characteristically of a class or of an individual" (Schurmann, 1973). It guides collective or individual behavior, sets the framework for diagnosing social ills, points the direction for social and environmental change, and often indicates effective solutions to social and environmental problems.

formation

How is ideology formed? Ideology comes from both private and collective thinking, from struggle for existence, or from struggle to change or maintain the world according to one's sense of "right" and "wrong." This struggle is often a manifestation of ideology, particularly when one stakes a claim to a position. Thus, ideology is not formed in a vacuum; it is created through multiple experiences and encounters with ideas. For some, an ideology may develop from experience as a member of a minority group, as a woman, or as a physically impaired person; for others, it may come from a deep-seated feeling of guilt about an issue and the desire to do something about it. For some, it comes from identification with parents or important people; for still others, an ideology is built from a variety of complex experiences. In any event, all people have ideologies. Some of them are well thought out, and some are not; some are conscious, and some are unconscious; some are practical, and some are not. Many things represent expressions of ideology: Beliefs in democracy, freedom, and justice; the way people relate to each other; the way people raise children; one's choice of foods or reading material; how one spends spare time; and how one relates to nature.

effect on behavior

In sum, an ideology is a manner of thinking that can provide a leverage for changing or maintaining peoples' behavior within institutions and within society. In some instances, an ideology is used to shape the political image of people to build cohesiveness. In one way, an ideology is a description of complex social phenomena that captures a vision of the future and one's imagination for action. In another way, it is more than mere description: It is used to immunize groups of people against diversionary tactics or to resist internal and external coercive pressure; it can stand as a signpost to guide collective behavior and becomes the touchstone for evaluating actions. Therefore, the value of any ideology depends on the effect it has upon behavior and upon one's very life.

THE IDEOLOGY
OF ENVIRONMENTAL ADVOCACY

The ideology of students and faculty in the EA Program at the University of Michigan attempts to integrate people on a continuum with other life forms. Human beings cannot afford to be totally anthropocentric; they cannot afford to act as though they are separate from their ecosystem or as though natural resources can be exploited with little regard to the environmental impact of human action. In the past, the mistake has been made of thinking that the natural environment and other life forms would somehow take care of themselves, even in the face of environmental disamenity. Because this anthropocentric world-view has dominated certain academic disciplines, most scholars have difficulty recognizing the significance of environmental problems and the severe constraints they may impose upon social, political, and economic institutions. In the 1970s and 1980s, a small group sociologists began to move away from the "Human Exemptionalism Paradigm" (HEP) and began to stress the importance of the biophysical environment as well (Canton and Dunlap, 1978 and 1980), (Buttel, 1976), (Buttel and Larson, 1980), (Morrison, 1980), (Schnaiberg, 1980) and others.

Commoner (1971) states, "the most important link between society and the ecosystem on which it depends is technology." Human-made technologies in many instances are in conflict with the ecosystem. In fact, beliefs in the infallibility of technology and the HEP have exacerbated environmental crises. Only now is one beginning to understand the interplay of economic forces, policy decisions, and resource management with technical environmental solutions and their long-term impacts. It is necessary to understand such interplay and incorporate it into the EA ideological framework. But one also needs to develop skills to be able to influence the decisionmaking process.

assumptions about society

In the EA Program, ideology concerning the environment is expressed in the assumptions ones makes about society. One assumes that wanton exploitation of human resources causes human suffering, degradation, alienation, despair, powerlessness, isolation, and self-estrangement. One assumes that power will accrue to those who already have it and not to those who are currently disenfranchised. One assumes certain forces present in this white-male-dominated society tend to keep minorities and women in chattel-like roles. One assumes that the ever-increasing use of nonrenewable resources in a high-energy and technological society will exacerbate the conflict between the "haves" and the "have-nots." One assumes that the by-products of a highly technological society will cause undue health problems and fouled air, polluted streams, and a scarred and unproductive earth. It is clear that such assumptions about society help determine one's diagnosis of environmental problems, set the context for exposing false mythologies and hypocrisies, and challenge competing ideologies that threaten society's very existence. Further, the problems faced are too complex to be limited to the diagnosis of highly specialized disciplines. To do so will give only a distorted view of reality and, thus, inadequate solutions for dealing with life-threatening problems.

visions and solutions

EA's ideology is also expressed in visions about and solutions to these problems. For instance, EA envisions people living more harmoniously with their environments in a world that accommodates a more equitable distribution of social and economic resources and mechanisms for the redress of social and environmental grievances. EA sees a society that reflects and supports a humanistic approach to life, devoid of the "isms" and the non-wanton exploitation of our highly cherished nonrenewable resources. EA supports a society where there is physical security and a basic protection against want, where there is plenty of food, clothing, and adequate housing for all. EA supports a society where socio-economic benefits and costs are spread as evenly as possible to everyone without distinctions based on biological characteristics. EA embraces a society where there is an opportunity to engage in meaningful, creative, healthy, and productive work in environments where people are proud of what they accomplish. EA supports a society where people have an opportunity to participate in making informed decisions that affect their personal lives in their work place, community, and nation. EA sees a society where personal growth and loving relationships are valued more than material incentives, where conflict is accepted as a natural part of the ongoing social fiber of the community, and where people may practice intentional nonviolence to press for just solutions (Gowan, Lakey, Moyer, and Taylor, 1976). These visions of and solutions to society's problems are kindled by EA's experiences, EA's knowledge of society, and EA's support for each other.

IDEOLOGY AND OGANIZATIONAL FORMS

An ideology is also a manner of thinking characteristic of an organization with predetermined goals and outcomes. If one agrees that organizations are rational instruments geared to do particular jobs, then the people who create and use them must do so on the basis of a particular set of well-integrated ideas. However abstract these ideas may be, they must have consequences for action. The more complex the organization becomes, the greater is its need for a systematic set of ideas to govern it, and the more dependent it is upon a systematic conception of unity. Thus, EA defines an organizational ideology as a systematic set of ideas with consequences for action that serves the purpose of creating and using an organization (Schurmann, 1973). In some organizations, ideology is not widely or consciously discussed, but may be reflected in a set of policies, rules, and regulations. Ideology also provides a way for an organization to care for its members that extend beyond policies or rules, embodied in an atmosphere of emotional support (or nonsupport) of individuals within the organization. An organizational ideology may be abstract or more concrete and easily understood.

reflected in the diagnosis

The ideology of EA is reflected in its diagnosis of organizations. Often organizations become stratified forms to support elite privileges, a condition that can lead to much conflict. Managers often attempt to control both the production and members of their organizations through a highly structured hierarchy, regardless of the impact that it has on the people within such organizations. Profit maximization becomes the primary focus, and employee needs and desires become subjugated to that end. Managers and owners of organizations feel that people come together for the same reasons--that is, for the "good" of the organization and its outcomes. What differences exist between people are seen to be secondary and can be resolved with minor changes, through well-defined grievance procedures, or by face-to-face interventions. Although these solutions are most attractive to managers, they have not always been workable. Such solutions are not always workable because people join organizations for various reasons, some of which are different from the purposes of management. People in organizations come from different cultural backgrounds, operate in different organizational roles, have varying interests and values, have a variety of traditions, are of different races and genders, and may be frustrated by a lack of fulfillment or self-expression. If one operates from the premise that these differences or problems are major sources of conflict rather than minor ones, then obviously major solutions and changes are required so that organizations will be more responsive to both individual and collective needs. If either the collective or any individual(s) feel dissatisfied or dehumanized and have few, if any, legitimate means for the redress of grievances, they may become highly politicized and drawn into conflict, or withdraw their emotional support from organizational activities.

questioning ideology

Withdrawal of emotional support may manifest itself in a variety of ways--from frequent absences to outbursts of anger and sabotage. In any event, EA students are trained to question the ideology on which different organizations shape their behavioral conduct and to question the role organizations often play in exploiting the people who work within them. Changing organizational forms and producing opportunities for personal and collective growth may increase both self-esteem and productivity and improve the quality of relations among the people who work in any organization.

ORGANIZATIONAL SKILLS

In the EA Program, a variety of skills are taught to help make the ideology operational. While many of these skills are value-neutral, they can be used to support the assumptions made about society. Therefore, the teaching of these skills is one of the most significant strengths of the EA Program. Key courses in the program reflect the assumptions made about society and solutions to them. EA also, however, tries to provide as many opportunities as possible for students to develop and practice skills outside the formal classroom. To this end, students are intricately engaged in a variety of social and environmental projects on and off campus, attend a variety of presentations and brown bag lunches where they collectively make decisions about the program and its future. They are encouraged to be activists while in school.

change arenas

In addition, alumni of the program are involved in building

both direct-action and support organizations in many social and environmental change arenas. Some of the direct-action activities EA graduates have been involved in to date are: (1) fighting rising utility rates; (2) protesting high property taxes; (3) stopping the building of expressways that destroy neighborhoods, relocate people, and take much-needed property off the tax roles; (4) resisting the building of nuclear power plants that are unsafe and have the power to destroy the planet; (5) taking action against industries which wantonly pollute and foul air and water and destroy land; (6) organizing political pressure groups against government incumbents to increase the allocation of monies needed for solar power research; (7) building alternative communities and cooperatives; (8) organizing better health care programs for citizens; (9) saving wilderness areas from development; (10) encouraging the development of national parks; and (11) supporting peace initiatives and demonstrations.

Other graduates of the program advise organizations by providing technical skills for organizational change, development, and direct-action strategies. They provide skills and support to direct-action organizations, helping them to develop hiring and firing procedures, organizational accountability, staff evaluations, affirmative action programs, long-range planning, and collective decision making. EA feels both direct-action strategies and organizational change and development are critical in building organizations that are strong and lasting.

organization building

Organization building often arises out of direct action struggles to help shape the direction of social and environmental change. When community groups have conflicts with antagonists over scarce resources or incompatible goals and values, community organizations can be built to garner forces to pressure antagonists or decision makers for more equitable distribution of resources and services. This kind of pressure has resulted in accomplishments ranging from obtaining a new stop sign at a local corner, to increasing welfare benefits, to even larger, more pervasive changes It is important that victories such as these, no matter how small, that result from direction-action be used to build strong organizational forms for groups to find self-expression for intentional social change.

Conflict used in organization building may be irrational or subjective conflict, i.e., the psychological phenomena of racial prejudice, stereotypes, and miscommunication. Organizations of all types have used racial prejudice and stereotypes to fend off assaults on white culture and to keep minorities and women in marginal, low-power positions. Additionally, community organizations have similarly used irrational conflicts against particular managers or executives of antagonist organizations to personalize issues in the media in order to attract potential members, thus increasing their clout. While we do not support the use of irrational conflict, its use should be linked where possible to understanding and supporting broad-based social change.

testing commitment

Because many people are seldom wedded to changing or building new organizational forms, EA's adversaries often have an easier time advancing their own ideologies. EA, however, not only has to build organizations from the ground up, but must also find ways to staff and maintain them over the long haul. This is no easy matter. Because social and environmental organizations often lack the ability to give large remunerative rewards, they are often faced with a highly transient staff. Because of low pay and high turnover, many of the newer staff often lack the necessary organizational skills and scientific background.

Thus, EA has the challenge to build or maintain organizations by turning inexperienced people--people who may be temporarily excited by an issue and are gone tomorrow, taking with them some of our time, energy, and skills--into experienced ones, committed and dependable over the long haul. It is important that EA test its students' commitment to help them gain clarity early on about their life choices, and then develop their skills and provide the resources necessary for them to work in a variety of organizations. The EA Program at the University of Michigan has done this, and has encouraged its graduates to network and share their experiences, support, and skills with one another.

ORGANIZATIONAL CHANGE, DEVELOPMENT SKILLS, AND IDEOLOGY

While students of EA in traditional organizations may find themselves in conflict with organizational ideology and want to change it, they would need considerable skill to alter existing organizational forms and attitudes rooted in traditional ideological norms and values. Explicit criticism of the organization and advocacy of collective decision making or new forms of accountability might meet with pervasive resistance in a variety of forms. When working within more traditional organizations, advocates have to gradually build new forms while being respectful of existing ideas and deep-seeded ideological values. In some instances, it is easier to build organizations from the bottom up rather than to change existing ones.

organizational hypocrisy

Organizations often reflect humanistic goals and objectives for society at large--that is, they want society as a whole to be more responsive to the collective needs of its citizenry or in someway to "right" a civil wrong that has been perpetrated upon various groups of people. But the internal workings and structure of organizations, which often emphasize ideological socialization to inhuman conditions or require subjugation of one's needs for the sake of the movement, are often oppressive. Considerable protest has been raised against the hypocrisies embedded in certain social change organizations. The conflict between external and internal ideology becomes paradoxical when organizational goals for society couched in humanistic terms are not granted to the organization's own members. Such organizational oppression, in a variety of forms and shapes, brings the EA commitment to social change into sharp focus and questions both organizational integrity and sense of purpose.

seeking cohesiveness

Historically, the style of community organizing advocated by Alinsky (1969 and 1972) is to build organizational power by engaging a constituency in conflict that will most likely yield successful outcomes. While celebrated victories have been the source of their internal cohesiveness and power, these organizations have avoided addressing deep-seeded internal problems. Specifically, they expect external victories to make members forget organizational inequities or problems. The ideology they champion for the poor, women, minorities, or the underdog is often inconsistent with how they relate to each other within the organization. Neither overt nor subtle forms of various "isms" should be tolerated in social and environmental change organizations--or in any organization. What EA champions for oppressed groups in society should also apply to the internal workings of EA organizations. Organizations that are inconsistent in this way are definitely not serious about social change.

CONFLICT; IDEOLOGY AND PROFESSIONALISM OR LEADERSHIP

In some instances, conflicts between the EA ideology and professionalism or leadership will exist. Professionals often develop an Olympian detachment of superiority. They accrue power unto themselves and become more concerned with protecting their own position than with being attentive to social change within or outside of the organization. Professional values are often also rather conservative and can be used to stifle democratic processes. Decisions by professionals are frequently handed down; they "know better" because they are experts and have been given the public trust. In some cases, this is undoubtedly true, but in other instances, one should question their decisions and the power they exercise within and outside their professional field.

balancing leadership

Conflict between the EA ideology and leadership will periodically come into sharp focus, too. In many organizations, one finds leadership positions distributed inequitably between males and females. Society's resources have been to a large extent allocated to develop patriarchal leadership. Males, therefore, have more opportunities to practice leadership skills than females, and they are usually expected to perform in certain institutionalized roles. Women are often defined as being less adequate than males, and their behavior is often prescribed beyond what is embodied in any job description to a point that is sexist in character. The same is true for minorities. Members of minority groups have little opportunity to practice leadership in integrated settings; they are often subjected to white cultural arrogance in predominantly white institutions and subjected to stereotypic views of racial inferiority. As both women and minorities continue to seek justice within organizations, considerable conflict will undoubtedly arise.

being true to the ideology

There are other potential tensions between the EA ideology and traditional leadership attitudes. Because leadership connotes efficiency, knowledge, and skill, people know who they can hold answerable for their problems; people often put their faith in leadership to make decisions about their lives. Yet, such centralized authority often breeds corruption, nonparticipation, and alienation among members of an organization. We know that professionalism and leadership are highly valued in American society, but questions about that value persist: How can one be professional and use expertise effectively without exploiting or dehumanizing others? How can leaders use their leadership skills without violating the sense of member participation or creating alienation and member dissatisfaction? How can one be true to one's ideology both internal and external to the organization?

Thus, it is important to find out how the organization treats women and minorities, where they are located in the overall structure, what part they play in collective decision making, and what provisions are made for their professional growth and development. Advocates have to be on guard for certain ideologies that are oppressive and stifle energy and creativity.

TWO ORGANIZATIONS WITH CONSISTENT IDEOLOGIES

Of course, some organizations attempt to have consistent ideologies. Following are descriptions of two excellent examples:

Illinois South Project

The Illinois South Project (ISP) was founded approximately 15 years ago, by graduates of the University of Michigan EA Program, to develop and organize multiple resistance to local coal companies that were responsible for buying and exploiting some of the most valuable agricultural land in Illinois. Coal company activities also threaten the future existence of the family farm and cause a number of problems related to subsidence, land reclamation, and property destruction. ISP built a collective structure to ensure a consistent internal ideology with its external mission. Important collective activities were: (1) weekly rotation of the coordinator of the organization, so that both men and women shared in this venture equally; (2) collective decision making at weekly staff meetings; and (3) a sense of trust, deep personal commitment, and caring for one another.

Over the years, ISP has gained a national reputation for many accomplishments in the state. Until recently, the quantity and quality of work done were related to the nonhierarchical organizational structure which motivated high staff participation and collective decision making. The members commitment to one another, the organizational structure, and the mission of holding coal companies accountable for their actions--which often had far-reaching environmental and social impacts--were more important to members than working elsewhere for higher wages.

While this collective structure survived for approximately 15 years, it eventually succumbed to outside pressures from funding sources which were not comfortable in subsidizing this collective work or felt that ISP's issues were no longer attractive enough to merit support.

Movement for a New Society

Until recently the Movement for a New Society (MNS) was another collective whose internal ideology seems consistent with the external one. Dedicated to developing a just world, this organization was based on the idea that a democratic movement must create structures for itself that support democratic group processes, personal and collective growth, political activism, macro-analysis of society, intentional nonviolence, and non-cooperation. While this counter-culture group placed a high premium on direct and nonviolent political action for changing society, the collective members took on outside part-time work as a means of supporting themselves and the organization. It was one of the few organizations that tried to model new lifestyles by acting on its ideology both within and outside the collective. But after more than two decades MNS, like ISP, was not able to survive in a hostile environment.

For organizations that have an affinity for building complex forms, the questions again become: How can this be done without sacrificing ideology? How can one build collective structures that will survive over the long run? There are no easy answers. EA suggests, however, that one try to make one's lifestyle within organizations consistent with the goals and objectives EA has for society.

SUGGESTIONS TO IMPROVE ORGANIZATIONAL FORMS

Three suggestions will hopefully strengthen large organizational forms and give currency to EA ideology:

ongoing study

First, EA feels that members in large organizations should become involved in ongoing study groups to increase their understanding of society and collective issues. Based upon this knowledge, collective members should discuss potential solutions and strategies for change, both internal and external to the organization. While time should be spent discussing how the organization can become more humane or productive, and how personal growth and development can occur among members, time should also be taken to evaluate the work performance of each member, including discussion of members' contributions to the study group and the overall mission of the organization. This should be done on an ongoing basis.

work study program

Second, a work study program should be instituted so that staff become intricately involved in working in other areas or in other roles within the organization. This technique opens lines of communication for relating across organizational roles and helps in developing greater sensitivity for those who have to perform different organizational tasks. This can be a powerful learning experience. EA also feels that organizational consistency between internal and external ideology lends itself to more productive members, sensitive to one another's needs. Because people feel they have control over their lives, they feel more motivated and lend emotional support to organizational goals and objectives. Both the Illinois Southern Project and the Movement for a New Society reported a deeply-felt investment by members in their organizations and witnessed high productivity and personal caring.

multiple issues

Third, EA feels the need for building multi-issue organizations which will seek basic structural changes in society. Building collective forms around a single issue is very likely to cause organizational entropy. Such organizations begin to falter because it is often difficult to maintain high energy involvement on a long-term bases in a single-issue organization, because people become bored or feel that victory is beyond their reach, or because the goal has been accomplished, leaving the group with little reason to exist. Therefore, it is important to consciously build multi-issue organizations so that issues can be taken on serially to help the organization exist over time. For example, issues and actions should be coordinated so that when the organization is beginning to lose motivation or energy around one issue, another one can be raised to bring a different set of supporters or constituencies to the forefront. This brings new people and resources to bear, providing new sources of energy for long-term survival. It is important to build organizational forms to last; time-honored social change is not a fly-by-night proposition.

summary

EA is different from other forms of advocacy in that it embraces an ideology that tries to improve the conditions of both people and their surrounding environment. It is neither anthropocentric nor biocentric in the extreme, but views life on Earth as a part of an ecological mosaic. People are not separate from nature, but rather are part of a continuum, with each part important to every other part. Over the years, the EA Program has been able to build a curriculum that not only speaks to these issues, but combines an ideology with skill development for building organizational structures to enhance social and environmental change. We feel that social change ideology should not only be extended to the broader community, but to people who work within organizations, dedicated to intentional social change. Moral integrity requires such an approach. EA is still in the process of becoming and changing, of trying to find effective ways of being powerful and intentional in its work. EA is excited and hopeful about the future.

CHAPTER 5

ACTION Research
and environmental
advocacy

Action Research (AR) refers to the systematic collection of information for purposes of facilitating social and environmental change. Its practitioners gather data to expose unjust practices or environmental dangers so that those community groups that are affected can make informed decisions.

EA, as indicated previously, is the critique of culture and political economy which provides direction for intentional social change by applying specific skills and strategies to solve social and environmental problems. EA's critique of both culture and political economy is integrally woven with AR. Such research embodies both a critique and a repetitive cycle of planning, action, observation, and reflection that consistently modifies thought, stimulating further learning and action (Carr and Kemmis, 1983). In short, this research is inseparable from EA's critique because its cyclical and integrative character gives both meaning and power to action.

To understand AR, one must understand the shortcomings of the traditional positivistic approach to scientific research. Even though this traditional approach is characterized as value neutral, free, and objective, a number of prestigious scientists admit that positivism, or the storybook account of research methodology, is more myth than reality. Moreover, governmental, corporate, and foundation funding of positivism has obscured AR as a legitimate research methodology. Even so, AR still provides an alternative research paradigm that empowers local people rather than detached researchers or outside agencies once removed.

AR is critical to EA because it empowers the critique of both culture and political economy, enhancing consciousness for undertaking meaningful action. It is a research paradigm that gains power because it is firmly grounded in contemporary problems and praxis rather than in attempts to prove or disprove predetermined theories or hypotheses. Thus, AR is most important to those interested in changing their own communities for social and environmental "betterment." This chapter critiques and discusses positivism, the forces giving rise to both positivism and AR, the meaning of AR, potential uses of AR, and a case study involving AR in Appalachia.

A CRITIQUE OF POSITIVISM

It is important that one understand positivism. It has dominated Western thought for centuries, dating back to such thinkers as Locke, Newton, Bacon, and others. It is an instrumental view of rationality--the view that human behavior can be understood objectively if one chooses sub-classes of subjects randomly for observation and if researchers remain value neutral and objective. Positivism assumes the only knowledge worth having is that which lends itself to quantitative analysis, contending that the only way to understand a phenomenon is to establish it scientifically (Robottom, 1983). It assumes that worthwhile knowledge can prove, extend, or disprove theoretical constructs. For positivists, the object of research is to understand a phenomenon that is determined by operations of physical laws and controllable boundaries (Carr and Kemmis, 1983). Problems involving values cannot be tested according to scientific methodology, and, therefore, they cannot be elevated to legitimate means of scientific inquiry.

revealing myths

Accounts that claim that social science research is value neutral, free, and objective are often more myth than reality. Objective scientific research is a myth that has permeated "scientific communities" from their inception and will probably continue to do so in the future. Dispelling this myth undoubtedly would be a threat to the scientific establishment because that could take from scientists their mantle as the high priests of knowledge and relegate them to the position of observable adjuncts of powerful interest groups. To dispel such a myth might lead taxpayers to withdraw their faith and support from the scientific community.

Mitroff (1974) writes an interesting account of how 40 top scientists participating in the Apollo lunar mission viewed scientific research. In a series of interviews over a three-year period, he demonstrated most vividly how value-neutral and detached scientific observations failed to exist among these scientists. He found, for example, "...that science does not advance through the single efforts of individuals, each dispassionately and logically testing their own ideas. Rather, it advances through a heated adversary process, which is fundamentally social, wherein one man tests his discoveries against the discoveries of another. Psychological energy and commitment infuses the whole process to such a degree that it is foolish to say that scientific inquiry naturally exhibits a clear-cut dividing line between individual scientists or between the contexts of discovery and of justification." He goes on to quote from interviews that demonstrate that science is more biased and political than value neutral and objective. Following are statements from several of his interviews with moon scientists:

* Bias has a role to play in science and it serves it well.
* I wouldn't like scientists to be without bias since a lot of the sides of the argument would never be presented.
* You don't consciously falsify evidence in science, but you put less priority on a piece of data that goes against you.
* The notion of the disinterested scientist is really a myth that deserves to be put to rest.
* One has to be deeply involved in order to do good work.
* I don't think we have good science because we have adversaries but that it is in the attempt to follow the creed and the ritual of scientific method that the scientist finds himself unconsciously thrust in the role of an adversary.

Obviously, these were no-run-of-the-mill scientists, but outstanding people in their fields. Perhaps these scientists felt secure enough in the scientific community to converse openly about biases found in the scientific research that shapes scientific knowledge. This does not mean, however, that because scientific information comes from an adversarial process steeped in biases, important contributions cannot be made; they have been made and will be in the future. The question is: Who controls the process and the content?

issues of integrity

We must also question the integrity of a science that passes itself off as value free and objective, and as the only legitimate method of generating worthwhile knowledge. Many methods for generating important information should have legitimacy. The scientific community should be more pluralistic, particularly in accepting other knowledge-generating methodologies.

Positivism has generally confined Western intellectual tradition to a single method of finding "worthwhile knowledge"-- the quantification of observation. This research methodology, which is based in the physical sciences, tends to restrict the focus of social scientists to short-run and isolated events. Thus, their research results often yield a limited range of meanings, creating an oversimplification of complex social phenomena (Smulyan, 1983).

Such research methodology poses some other problems related to the treatment of subjects. While paternalism and social control of the "helpees" underlie most traditional research studies (Pine, 1981), such research is not committed to egalitarianism, collaborative enlightenment, or consciousness-raising to increase potentials for self-affirmation and struggle. Embodied in positivistic research is a dualism that manipulates, objectifies, and sets subjects apart from the aggregate as random samples to be studied for the sake of increasing the power of elite groups. This form of knowledge generation not only shapes society's worldview and determines relations with others and the environment, but it has set society on a perilous trend: One of objectifying humans; one of exploiting natural resources in the best interests of the few rather than the many. Capitalism flourishes through positivistic approaches that can be used to justify the objectification and exploitation of natural resources. Although many relish affluence and conspicuous consumption afforded by capitalistic enterprise, the by-products of capitalist production often threaten health, the environment, and even lives.

the random sample

Why is the random sample so important in positivistic research? Extrapolating research findings to a larger aggregate for government agencies and corporations to make critical decisions to augment their control, markets, and profits, requires observation and analysis of random samples. In positivistic research, those generating the information give up the locus of control of that knowledge to scientists, bureaucrats, or agencies, thus empowering them to make decisions about larger classes of subjects. Apart from the use of positivism, scientific "biases" enter the picture in other ways, too. Those funding the research often determine or influence the topics to be investigated and the extent of the investigation, by allocating or withdrawing remunerative rewards. Thousands of foundations and millions of dollars each year fund special research projects that serve the immediate or long-term interests of the funders and elite groups.

This does pose an interesting question. Should powerful interest groups determine the character and direction of research in institutions of higher education? To be competitive, grow, and develop, research institutions have given up considerable control of knowledge generation to outside interest groups, in turn helping those groups further their own particular needs. To enhance the power and influence of elite groups, significant numbers of social scientists focus their research observations upon the plight of the less fortunate and very little upon the power and influence of the elite groups themselves. While such research outcomes often blame the victim for consumer waste, their lack of emotional support of recycling projects, their impoverished or detestable condition, or their inability to achieve, little research focuses on the structural or systemic factors thwarting meaningful action to enhance their well-being. Research in universities not only serves elite interests, but much of the research done is subsidized by taxpayers support to universities, making the very role of research institutions questionable. EA hopes that this scientific discourse will move to center stage as taxpayers become more aware of research subsidies to the private sector.

FORCES THAT GIVE RISE TO CONTEMPORARY POSITIVISM

What are forces that have boosted positivism in modern times? How has positivism obscured other research approaches?

in reaction to Sputnik

A good place to begin is with the Soviet launching of Sputnik in the late 1950s, an act viewed as threatening to U.S. world hegemony. This feat had both scientific and political ramifications. Suddenly, public schools and institutions of higher education needed help in training new generations of science students. America had apparently been derelict in its educational responsibilities. To overtake the Soviets, John F. Kennedy's "New Frontier" and space exploration were given high priority. It took approximately ten years to reach the moon--a feat unparalleled by any previous event in scientific history. To eradicate poverty in America, Lyndon B. Johnson's "Great Society" program, which allocated millions of dollars to communities across the country, was implemented. Money was also given to social scientists to study conditions of poverty and to extend social science theories. More recently, Ronald Reagan's "Star Wars" program and other military build-ups have provided untold amounts of money for positivistic research endeavors.

to stray is frowned upon

Few college and university courses focus on nonpositivistic research methods. Faculty are usually rewarded for being detached, value neutral, and objective in using random samples and complex statistical models to quantify data. To stray from positivism is frowned upon in academic circles, and thus so is AR.

FORCES THAT GIVE RISE TO CONTEMPORARY ACTION RESEARCH

Yet, as the perceived threat of Sputnik and the war on poverty brought a call for spending millions of dollars on positivistic research and development, another phenomenon was also taking place. Some of the same forces that gave rise to positivism also gave rise to AR.

accountability movement

The "accountability movement" surfaced as an important political event. When the rise of professional power encroached upon decisions traditionally made by individuals, the civil rights, women's, peace, and environmental movements emerged as attempts to make government and professional groups more accountable. Mistrust of government officials arose over the ethics and legality of the War in Vietnam. Mistrust of government officials and corporate executives resulted from their lack of commitment to racial and sexual justice and an environmentally sustainable future. The Watergate affair even further eroded public confidence and trust in government. Currently, one witnesses an unprecedented concentration of corporate power through corporate mergers and takeovers. As this concentration of economic power takes place, the Reagan and Bush Administrations, through the Supreme Court, have attacked the rights of minorities, women, and the handicapped, while also relaxing environmental regulations to leave communities feeling quite unprotected.

taking control

These circumstances would seem to foster conditions for people to want to take control their lives. Some are no longer content to leave critical decisions to corporate interests. Distrust of government and corporate officialdom has given greater impetus to the resurgence of AR as an important and legitimate tool. Undoubtedly, there is a clear need for a research paradigm to enhance greater participation of people in making informed and meaningful decisions about events affecting their personal and community lives. Yet, with respect to gaining legitimate status in academia equal to that of positivism, AR has a long way to go.

ACTION RESEARCH

In contrast to positivism's bias and use by the establishment for control, profit, and exploitation, AR provides a conceptual framework for analysis of such realities. Even more importantly, it has merit in its own right. The following discussion attempts to raise critical points related to AR, setting the stage for a dialogue concerning its potential and efficacy. Included will be one case study, showing how action research was used in the furtherance of partisan efforts to achieve community goals.

what AR has to offer

AR has much to offer community groups in general. Obviously, this type of research breaks with the confines of political conformity and intellectual dispositions by exposing false mythologies, old values, and hypotheses that favor elite groups. By the same token, of course, AR challenges the traditional position that reserves significant research for the well-qualified professional. Community groups can also do high-quality research, as evidenced by the example to be discussed. Community groups are seldom opposed to professional researchers; they should, however, be opposed to the role these experts often play. Often the emphasis upon professional expertise changes the rules of exercising community power; problems of political choice become buried in debates among experts over highly technical alternatives (Peattie, 1968).

Technical information can be understood if professionals take the time to explain results in lay terms. When scientific processes and decision making exclude community participation, the democratic process is usurped. Even though feeding positivistic research results into the policy arena may be more efficient than offering it for democratic scrutiny, the process circumvents community input. In the end, policy decisions may never be accepted unless community groups are involved in research and decision-making processes.

Community groups can even the odds somewhat with AR. With this tool, they can empower themselves as a political force to be reckoned with. Many community groups have researched issues of concern, using their newly found information for collective action. All too often, AR is frowned upon by outside agencies because it "fails" scientific standards. It is often viewed as nonprofessional or non-scientific research and, therefore, less than legitimate. AR, nonetheless, has much to offer scholars and lay people alike.

a different epistemology

AR requires a different epistemology. First, the complexity of the social world and changes over time, as well as the existence of cultural differences, would make it impossible to discover social or cultural "laws" comparable to physical ones (Hodges, 1944). Second, the human experience must be understood historically. AR cannot achieve the ideal of absolute detachment, because it survives and becomes critical within a social and historical process. The historical process bounds the experience in such a way that the action researcher is an extension of that history and all its traditions, values, and norms, which predetermine conditions for viewing society and acting upon it.

Action researchers are not detached and noninvolved in the generation of knowledge, but they recognize more than positivists that their circumstances, language, customs, and values cannot be separated from the phenomena under study or from their own praxis. The articulation of their analysis comes from praxis that is the result of struggle and political discourse; so does practical truth.

use of praxis

In AR, truth of statements and information is confirmed and evaluated through practical discourse, planning, action, observation, and reflection, using each step of the repetitive planning cycle to modify events as indicated by praxis. Smith (1983) states that "truth can only be a socially and historically conditioned agreement; what is true is what we can agree upon at any particular time and place." Individuals and groups determine their own reality based upon relevant social and cultural interests. In this context, the criteria for truth involves authentic insights grounded in the community's own circumstances and experience--an ever-evolving dialectical process that is not guided by predetermined theories or constructs. Truth and AR processes are one and the same. The research processes themselves do not take place in linear fashion, but rather in dialectical and cyclical progressions, moving from specific micro-contexts to macro-contexts and back once again to anchor one's truths or social statements in the field (Cave, 1982).

In AR, truth is evaluated according to its usefulness in furthering ideas to authenticate strategies for social change. The issue here is not "ultimate" or "universal" truth, but a process of discovery of ideas and strategies embedded in the historical process, which may actually change over time as complex societal relations, interests, and needs change or as new data become available. Truth is relative to the experiences and circumstances of those engaged in praxis, making actors alone the final arbiters of truth and of data interpretation. Rules, principles, and theories do not suffice (Carr and Kemmis, 1983).

Rigor, like truth, comes from the logical and political coherence of interpretation and practical discourse following moments of action and self-reflection. Rigor comes from planning, based upon critical discourse and repetition of the planning cycle (Carr and Kemmis, 1983). Both praxis and monitoring actions, as well as their consequences, refine truth and rigor. Rigor arises from compelling analysis of social events and increases precision through praxis.

political discourse

Action research empowers people in an organization to make their political discourse with antagonists or target groups more symmetrical in character. Without AR, power relations with antagonists may be asymmetrical, relegating community groups to positions of political disadvantage. AR can be used to build powerful organizations to make meaningful demands upon antagonists, making the system more responsive to the collective social and environmental needs of the community. At a more general level, its purpose is to confront ideology that is used to reproduce oppressive rather than liberating social relations. Its ultimate purposes are: to intervene critically in all patterns of action which fragment communities, destroy the most cherished resources, and pollute the environment; to intervene critically into societal practices that perpetuate myths and misunderstandings; and to work to increase principles of environmentally sound values, behavior, and the principles of economic democracy.

AR is a relentless critique of culture and political economy, both of which otherwise maintain social and environmental irrationality, social injustice, social fragmentation, and the coercive domination of both human and natural resources. AR serves to identify those aspects of the existing order that thwart and frustrate intentional social change. Obviously, praxis or AR is political, empowering people to take meaningful action to liberate themselves from the shackles of oppression, alienation, and the established order that thwart and resist meaningful change.

demystifying research

One of many roles of EA is to demystify research by showing how such research is not reserved for detached academicians, but can be performed by anyone--including those concerned with intentional social change for their own protection against dominance and exploitation or the protection of their environment against wanton destruction. Communities can no longer afford to depend solely on government, corporate, or university researchers or decision makers to solve problems of oppression and environmental degradation. Increasingly, communities must depend on themselves.

As technology abounds, so do toxic dumpsites, acid rain, pollution of surface and underground waters, radioactive contamination, and, on the social side, high unemployment, increased welfare dependency, crime, and sex and race discrimination. It is critical that communities not leave these problems to be solved just by outside agencies or researchers. Unfortunately, there are few rewards for AR in academia. As a result, few participate in such endeavors without great professional costs. Perhaps communities in some way can help by supporting and rewarding academic involvement in AR.

ACTION RESEARCH
IN THE APPALACHIAN
REGION

The following case study fails to meet the requirements of AR in its purest form, but it is the kind of research that should be supported and encouraged where possible. A two-year study was carried out by a group of community organizations, scholars, and individuals associated with Appalachian Alliance, a coalition formed after the 1977 floods in that region. This study differed from most studies of the region in that it grew out of the needs of indigenous people and was carried out by the people themselves, with the help of various university and non-university sources. The Appalachian Regional Commission and private foundations supported this collective research endeavor.

Both training and demystification of research helped 65 individuals and their numerous supporters collect information from 80 county court houses. They retrieved information from tax rolls and county deed books, compiling data on 55,000 parcels of land--approximately 20 million acres of surface and mineral properties. They compiled over 1,800 pages of data in a regional volume, and six other volumes were compiled for West Virginia, Virginia, Kentucky, Tennessee, North Carolina, and Alabama. The volume for each state contained a detailed profile of land ownership. More specifically, these data included the top ten landowners, the top ten mineral owners, the taxes they paid, and whether or not they lived in the county. Also detailed were the taxes paid by absentee landlords and corporations, comparing them to taxes paid by local residents.

fundamental issues

Previous studies, of a more traditional nature, had been confined to standard sociological categories: population growth, manufacturing, tourism, religion, social welfare, health care, and folklore. They had ignored the fundamental issues of natural resources and land ownership. The political economy of land holdings received little academic attention, if any, until Harry Caudill's book *Night Comes to the Cumberlands* was published. This book helped provide the impetus for a community-based research endeavor.

The results of the study clearly showed that property, in general, was underassessed. Schools, roads, and local revenues were thus underfinanced, with the burden of taxes shouldered not by large landowners, but by home, trailer, and automobile owners. By bringing rural and coal land assessment up to par with houses and automobiles, monies could be allocated for increased quality of services for poverty-stricken areas.

putting AR to work

In Kentucky, for instance, community groups are using information from the Kentucky Land Ownership Study to challenge underassessment of property taxes paid by corporate and absentee landowners. The Concerned Citizens of Martin County (CCMC) is working with other such groups to enforce state-mandated increases in property assessments. The landowner study provides concrete figures to back up CCMC's contention of widespread underassessment leading to abject poverty. The land study provides residents in the area with facts and figures about the local political economy and how the tax structure benefits some and not others. The land study makes it easier to understand why there is so much poverty in the midst of a land richly endowed with natural resources.

This land study shows how communities and universities can

work together to empower those most affected by underassessment of corporations and absentee landowners. Neither predetermined theories nor extensions of such theories were major considerations for the study. Even though researchers and community people may not have gone through all the processes of observation, planning, action, and reflection, in that order, they were undoubtedly able to use the AR paradigm to empower themselves to change an inequitable property tax structure. This research project also displayed how both academic and community people can engage constructively to improve conditions of indigenous people. To this end, AR should be given more legitimacy within universities. AR is a way for universities to form partnerships with community groups, leading to democratization of knowledge and empowerment of such groups to make critical decisions about their lives. If research in its most "rigorous" methodology can still be political or partisan, as indicated previously, then it benefits universities and funding sources to support more pluralistic approaches to research.

Unfortunately, because of the ways that universities and colleges are tied to our political economy, there seems to be little support for increasing AR significantly in the near future. It is, therefore, left to individual faculty members to build support systems within and across university lines to provide AR technology to help empower community groups.

questions of impact

Environmental advocates, like moon scientists, view positivistic research as biased and political. Yet, these two groups are radically different, in that moon scientists are production-oriented and environmental advocates are impact-oriented, and each makes different assumptions about social change and society. EA is concerned with the impact of technology upon physical and environmental well-being and on our social and political institutions as a whole. Thus, AR can make important contributions. AR not only helps one understand the political character of positivism, but it empowers the critique of culture and of political economy and the ability to act upon that critique. The AR cycle provides a "curriculum" for citizen groups or advocates, engaging in meaningful action. Moreover, AR has the potential of being a formidable countervailing force against powerful interest groups and traditional cultural patterns that reinforce social injustice and environmental degradation. AR must continue to be used for justice and peace throughout the world. It can be done.

CHAPTER 6

THE TEACHING of environmental advocacy: assumptions, issues, and strategies

Just as AR challenges traditional research paradigms, the teaching of EA challenges well-established traditions of pedagogy by viewing students as resources and co-facilitators with faculty in the learning process. Although this teaching and learning process is different, it is not the intent here to argue that it is the best method for teaching accumulated knowledge; other methods of teaching clearly have legitimacy as well. It is the intent here, however, to list assumptions made about the creation and use of knowledge and teaching which form a basis for criticizing biases inherent in traditional epistemology. The teaching of EA not only critiques the knowledge base and the process of education, but also critiques lifestyles, institutions, and society as a whole. One advocacy seminar, entitled *Social Change, Energy, and Land Ethics* (SCELE), actually attempts to empower students through knowledge and egalitarian principles.

GENERAL ASSUMPTIONS

Before describing the characteristics of that seminar, however, it is important to consider a list of general assumptions about the generation and uses of knowledge:

assumption 1:
knowledge is
a monopoly of the elite

Because elite groups make remunerative rewards available to universities, colleges, and research institutions to generate knowledge in their behalf, they stand to gain a lot from scientific breakthroughs which increase their profits or their control over more efficient production. They stand to gain from the infrastructures of these institutions of higher learning, even though the institutions are supported by public funds. Obviously, specialized interest groups, even though they pay overhead costs for funding research projects, would have to pay considerably more if they had to support more of the true costs of the university infrastructure. Narrowly defined and powerful interest groups become consumers of this knowledge, thereby increasing their monopoly control and profits by using scientific breakthroughs to increase their efficiency in exploiting both human and natural resources.

assumption 2:
knowledge is
guarded by professionals

Often-fragmented and specialized knowledge is generated for not only elite groups, but also for professionals who jealously guard and monopolize scientific knowledge, giving them considerable power that threatens democratic decision making. Unfortunately, the specialized jargon found in professional journals excludes the general public and prevents that public from making informed decisions. Because such knowledge has a built-in bias, serving the interests of elite groups and professionals more than other groups in society, and because it is specialized and fragmented, it helps create narrowly defined visions that may differ from macro or holistic perspectives of society.

assumption 3:
knowledge
can be violent

Society has always valued knowledge that allows one to coerce the world into meeting stated needs, regardless of its inherent violence (Palmer, 1983). Both the direct application of knowledge as well as its by-products may cause social and environmental problems. Some examples are: The clear-cutting of forests for timber has destroyed rain forests, causing multiple soil erosion problems; Acid rain destroys thousands of lakes and forests not only in the U.S. and Canada, but throughout the world; The contamination of underground water supplies continues to grow at an alarming rate, particularly in the U.S.; Thousands of toxic dumpsites across the country threaten water supplies, lowering property values and causing serious health problems to those living in close proximity.

assumption 4:
knowledge
can reduce jobs

In other instances, scientific knowledge is used to create new technology, such as computer driven machines, or it is used by elite groups or government to enhance power and control over markets and labor to increase profit margins. Shaiken (1980) states that the burgeoning influence of high technology not only threatens jobs in traditional areas, but also in areas that promised jobs for those displaced by the manufacturing sector. A net loss of jobs, due to capital-intensive technology, will definitely create more control over labor by management, as workers compete for scarce job opportunities.

assumption 5:
knowledge is
seldom questioned

Because students are seldom encouraged to question the authority role knowledge plays in course content and in their everyday lives, they seldom question the biases inherent in knowledge. The accumulation of scientific knowledge is important, but one must question its wholesale character and meaning.

assumption 6:
the traditional lecture
has real limits

It is assumed that lectures are the most efficient way of disseminating accumulated information, mainly because it has been done this way throughout the centuries, and because it is assumed that students will give most of their attention to the lecturer. However, current research clearly indicates that college students at best can only concentrate 15 minutes at a time (CRLT, 1978). This is not to say that lectures are not important, but that students might be better served if the 50-minute lecture format were divided into two 15-minute periods, each followed by discussion. Such a format would be more consistent with students' concentration stamina. Also, short discussion following the lecture allows students time to integrate the subject matter more fully.

assumption 7:
students can be
sources of knowledge, too

Although students are seldom seen as viable information resources to be taken seriously by faculty, they nonetheless are and can become important sources of information if given opportunities to rely upon their own resources. It is not uncommon for students to see themselves as reservoirs of important information; it is not uncommon for them to help each other to extend their learning far above and beyond the requirements of the course. Class discussion, utilizing student resources, can be just as motivating as lectures and in many cases even more so.

assumption 8:
grades do not measure
all knowledge learned

Traditional teaching assumes that grades reflect both quantity and quality of student learning. An "A" grade indicates that a student has learned more than another student with a "B." Perhaps, however, it reflects that the "A" student learned what was important to the teacher, even though the learning process may extend well beyond the boundaries of examinations, and therefore of grades. How does one measure information learned by students that is not on the test? It can be done, but such evaluation may prove less attractive, since it is far more time consuming.

assumption 9:
grading on a curve
is problematic

The efficacy of grading on a curve is another problematic assumption made about knowledge. In a given semester, 85 points may factor out to a "B," while the following semester the same number of points may factor out to an "A." Does this mean that the latter students learned more than the former ones? Not likely. If a student were in class with high achievers during a given semester, that student might experience a lower grade than in a class with low achievers. Thus, the extent of learning that is reflected in grading on a curve is by chance, much of which depends upon the academic composition of the class during a given semester.

other concerns

Other concerns about teaching go beyond the basic assumptions articulated above, and faculty over the years have tried various innovative approaches to teaching; but lectures are still the most common method of teaching and will probably continue to be so, even though they may have become less important since the printing press, electronic media, and computers. The technological extension of brain power through the instantaneous informational feedback from around the world enables everyone to be less dependent on traditional forms of instruction.

traditional teaching

Perhaps a heavy reliance upon lecture fosters authority-dependent relations between students and faculty, thus establishing asymmetrical relations, creating social distance between them. Students become peripheral spectators in a drama where the teacher is the actor (Palmer, 1983), and they are socialized to be passive-dependent as teachers deposit information into their receptacles (Freire, 1974). This authority-dependent relation fails to liberate or engage students in critical thinking which could empower them to take charge of their learning. In fact, authority-dependent relations in infancy, latency, adolescence, and early adulthood become so ingrained that students often feel less educated without lectures.

SEMINAR DESIGN AND REQUIREMENTS

During the first four weeks of the seminar *Social Change, Energy, and Land Ethics*, students start the process of researching and designing their own seminar based on predetermined topics selected during the first two classes periods. The faculty, for the first four sessions, provide readings and facilitate discussion on such topics as "Theory and Critique of Social Class in Modern Society," "Understanding of U.S. Race Relations within the Context of Natural Resource Distribution," " Theory and Critique of Sexual Oppression in Modern Society," and "Resource Structural Dependency and Geo-Politics."

In the past, students have worked in pairs to research and retrieve information on topics such as "Bio-Technology and Its Implications for Food Production and Processing in the United States," "The Family Farm Crisis and its Impact upon Rural America," "Advantages and Disadvantages of East/West Coal as Sources of Energy," "New and Renewable Sources of Energy," "The Nuclear Arms Race and its Impact and Potential Impact upon Humanity," and "Nuclear Power, Economics, Safety and its Relation to the Arms Race." Though these topics are usually chosen by the faculty, and may differ in their currency or importance, students may choose their own related topics.

In any case, students are responsible for designing a seminar around them. This includes typing up a bibliography for class distribution and organizing and duplicating readings for their peers, to be handed out one week in advance of their seminar. Students are responsible for facilitating a three-hour seminar and for designing a two-hour laboratory experience. While the seminar has a predetermined format which is discussed later, the laboratory experience has no predetermined structure, allowing students to be creative in relating additional information to their topic. Some have used role plays, simulations, movies, outside speakers, and the like.

seminar structure

The structure of the seminar is as follows:

10 minutes--Excitement sharing: This is a time for participants to make one or two statements about an experience that was exciting. This may be either small or important changes in their lives, important news stories, or something they heard or saw on television.

10 minutes--Announcements: There are lots of exciting events on campus or in the community that people usually share with each other.

10 minutes--Selection of articles: Students select articles or readings for the following week.

45 minutes--Seminar reports begin: Each report is a total of 10 minutes. People should relate three or four main points from the readings. Additional points are integrated during group discussion. After each report, participants may

ask questions for clarification. (During both the report and the discussion, one of the facilitators listens for and records on newsprint problems, potential action projects, and strategies for change.)

15 minutes--Break: At this point, those responsible for the seminar are to serve snacks. Food is important.

60 minutes--Group discussion: Participants in the seminar may discuss facts and ideas that failed to surface in reports or ask additional questions regarding clarification. Also during this time, the recorder writes down potential social change strategies to be reviewed at the end of the seminar.

30 minutes--Strategy session: Students brainstorm strategies for change and discuss their feasibility.

10 minutes--Evaluation: At the end of each seminar, there is an evaluation to help students become more effective seminar participants. This is the time when students focus upon the group process and strengths and weaknesses of the seminar itself.

grading

In the course *Social Change, Energy, and Land Ethics*, students would automatically receive a "B" grade for the seminar if they attended, participated, organized a seminar discussion, and designed a laboratory experience. They had to contract with the faculty for extra work to receive a grade higher than "B." In this seminar, an attempt was made to get students to work for knowledge rather than for grades; grades sometimes stand in the way of learning and meaningful student-faculty interactions.

When students contracted for grades higher than a "B," they took even greater charge of their learning and engaged in activities that extended far beyond the requirements of the course. Such activities are usually an empowering experience. One student, for example, facilitated a seminar on food production and processing, from which she eventually wrote a book. Another student recorded a radio tape on nuclear power plants, showing their dangers and encouraging people to participate in the Seabrook Occupation. *The Ann Arbor News*, the local paper, gave another student 15 column inches for his article on wood fuel as an energy alternative. Two students' seminar on food eventually became a course for the University of Michigan Extension Service. Four students from the seminar and a couple of others who did not participate in the seminar formed the Ann Arbor Collective to write an energy policy statement which was presented at the Non-Government Forum of New and Renewable Sources of Energy in Nairobi, Kenya. A group of six students raised $5,000 to organize a conference on robotics and high technology that was attended by 700 people from labor, industry, and the university community. Although not all students engaged in such elaborate extracurricular projects for grades higher than a "B," in most instances they do considerable work of a creative nature.

logs and journals

Another important component of the learning process is reflecting and integrating knowledge through logs or journals. By writing logs, students are encouraged to organize information in ways that clearly show that they understood readings and seminar discussion. At first, students usually have difficulty writing quality logs. Most have never done this before. But as the seminar advances and they gain more confidence in reflecting on their thoughts and writing them on paper, they are able to improve both their logs and learning substantially.

Logs help students become less guarded in group discussions. When students prepare for active group participation by writing and reflecting on readings, this, along with the design of the seminar, helps break down barriers of communication among students and between students and faculty. Logs improve the quality of discourse, encouraging students to raise opposing opinions which, in turn, improve quality thinking. In some cases, participants are able to change previous firmly-held ideas, opinions, and attitudes through such discourse.

contemplation

Near the end of the semester, at least one three-hour block of time within the seminar is devoted to "clearness" for the future. Many of these ideas in this activity come from "clearness meetings" used by the Movement of New Society and by Palmer (1983). All students, following the completion of the seminar they personally facilitated and near the end of the semester, are required to contemplate and write down what they see themselves doing in the five years following graduation. They are asked to write down such assets as personal work experience, course work completed, personal and organizational skills and resources, and skills they need for the future. Each student has about 20 minutes to present their view of the future to seminar participants. This is done in small groups of four or five persons each.This is followed by questions for clarification. This activity is designed to help students gain clarity on their future and what they need in order to get there. Students in supporting roles may ask such questions as: Is your plan for the future consistent with your feelings about social change? Are your social change goals for the future realistic? Do you need additional skills and resources to get to where you want to go?

Students are forced to integrate their learnings and resources with life goals. Through a "clearness seminar" students are helped, by the questioning format, to re-evaluate old and sometimes new assumptions or premises, sharpening their vision about the future and ways of getting there. The session ends with students giving feedback to presenters and vice versa. This feedback is both positive and constructive; it is designed to affirm one's existence, supporting one's life-goals and objectives. A sense of empowerment comes from feedback; it helps liberate as well as increase affinity relations among students and faculty.

group process

Both substantive knowledge and attention to group process also increase affinity among students. Affinity group building is started at the beginning of the semester; students and faculty spend a whole day together (usually the first Saturday following student registration), sharing personal and professional experiences. They not only use this time to learn about small groups, but also to engage in conversation about course expectations. Equally important is a three-day field trip to southern Illinois, during which the group camps, sleeps, prepares food and eats, plays games, and sings songs together around the campfire. The trip involves visits to stripmines and talks with miners, coal executives, state legislators, community people, and farmers to obtain information on the social and economic impact of coal development in the region. Camping together for two days is one of the highlights of the course; it helps to build closeness among students and faculty.

evaluation

It is important to build affinity with one another to enhance learning. Evaluations are important because cerebral activities are de-emphasized, providing rare opportunities for students to express their feelings regarding the seminar. Yet, when students are confronted by readings or by their peers on life-long values-- values which may provide guidelines for feeling, thinking, and perceiving the world--such a process can be rather painful. Once these value systems are shattered, then new ones or alterations of old ones must come into play. Some require patience as they take on new values and new ways of feeling and thinking about themselves and others and the world in which they live. Through affinity, some students have developed friendships that will last a life time.

Because evaluations are critical to the learning process, a one-time evaluation at the end of the semester, although important, is definitely too late for any meaningful changes to take place. Therefore, ongoing 10-minute evaluations at the end of each seminar are solicited by the group to generate data for purposes of improving the seminar. The information generated from each evaluation is acted upon in the next seminar whenever possible.

ASUMPTIONS UNDERLYING THE TEACHING OF EA

As should be clear from the foregoing, assumptions underlying the teaching of EA differ from those of traditional pedagogy. The teacher is not the locus of control within the EA classroom, but rather a co-facilitator with students in the learning process. Authority-dependent relations are thus discouraged, enhancing independence of thought, personal autonomy, critical thinking, and quality interaction with peers (Abercrombie, 1970) and faculty. The classroom is student-centered, emphasizing student participation and symmetrical relations between students and faculty, fostering self-affirmation and empowerment. This creates an atmosphere for students to take charge of their own learning. Through this process, they liberate themselves from the shackles of authority-dependent relations by becoming actors, not spectators, and by coming together in human solidarity, creating a belief that they can transform the world in which they live.

Active participation in learning requires students to make perceptual shifts from being the object of knowledge to being the source of knowledge, from being authority-dependent to being self-empowered, and from being passive participants to being active learners and teachers (Abercrombie, 1970). Group participation helps students share their knowledge with peers, preparing them to be open-minded in teaching and learning throughout life.

locus of control

Although it is difficult to pass the locus of control from faculty to students, it is necessary for learning for empowerment. Captive audiences or students as spectators can easily reinforce the needs of faculty for status and recognition. Letting go of power is difficult for faculty because they may come to be viewed contemptuously by their colleagues or, in some cases, by students for failing to provide the presumed necessary "rigor" or "discipline." This will undoubtedly always be a problem.

One way to foster democratic principles of egalitarianism is to have students read different articles for class discussion, allowing opportunities for each of them to make a unique contribution. This, in turn, helps them make the perceptual shift from being the object of knowledge to being the source of knowledge, shifting the locus of control from the teacher to the group as students become both teachers and learners. Democratic egalitarianism is further based on analysis and discussion of weekly readings in the seminar. If seminar participants shun this responsibility, they not only disadvantage themselves, but they undermine the principle of participatory democracy.

equality of participation

Students are encouraged to participate equally and to cherish the values of democratic egalitarianism, but sometimes some will tend to participate to dominate. Questions or discussion may not come from one's desire to know or from critical thinking, but from competitiveness and a struggle for power and influence over peers. Palmer(1983) sheds light on this:

> I have been in classrooms where people seemed to be pressing each other, asking hard questions, stripping off the veils of falsehood and illusion. But behind the appearances, something else was often going on. In an inhospitable classroom, many questions do not come out of honest not knowing. They are rhetorical or political questions, designed to score points with the teacher or against the students, questions asked not for truth's sake but for the sake of winning. In such a setting it is nearly impossible to reveal genuine ignorance--which means that genuine openness to learning is nearly impossible as well.

Although rhetorical or political questions may be designed to score points with the teacher, there may also be multiple forms of resistance to authority when students and teachers are involved in asymmetrical pedagogical relations. Repetitive questions may be more than merely questions, and indicate passive-aggressive behavior against authority. Such behavior may also take the form of handing in assignments late or class absenteeism, or both. In addition, students who say little in fast-paced discussion may have deep insights into the subject matter and much to contribute if they are allowed to participate. Under such unfortunate conditions, authority-dependent relations in classrooms, student competitiveness, and dominance are seldom dealt with in any meaningful way. In the EA seminar, these problems are dealt with through evaluation, usually at the end of each seminar, during which time it is discussed how one can improve relations with one another in the context of transferring substantive information.

personal change

Personal change is important, too. Everyone must start first by asking, "What personal risks am I willing to take? How do we go about acting out our commitments to ourselves and to one another?" Relating to others only on the basis of past experiences or the way one knows best is often not good enough; one must also understand others' origins and culturally determined patterns of oppression and dehumanization. Some cultural patterns are acted out in the seminar with respect to the "isms" and culturally defined arrogance that dehumanizes people in an attempt to render them less important. This calls for increased understanding of who the students are and where they come from, thus helping to liberate everyone from such patterns so that they can engage in more meaningful cross-sexual and cross-racial interactions and fundamental social change. That is why the discussion of readings on social class, race, and sex selected by the faculty at the beginning of the seminar is so important.

It is important for faculty to share not only their knowledge but also their values with students. Too often, the apparent value-neutrality and objectivity of faculty is mere pretense. Faculty are often quite biased in their presentation of some theories to the exclusion of others. In EA it is felt that research is biased because it is bounded by discipline or narrowly defined theories. Faculty should be honest about values, ideology, and the limitations of narrowly defined and fragmented knowledge so students can consciously choose to accept or reject value-latent information.

TEACHING PRAXIS-- ACTION AND REFLECTION

Meaningful knowledge is a base of power that allows people to make important choices regarding their personal lives and the environment in which they live. Inaccessibility of knowledge may create conditions of powerlessness. Both power and powerlessness corrupt. Too much power in the hands of elite groups often encourages them to make decisions in their own narrowly defined interests rather than in the interests of the "collective good." One can see the abuse of power in the amount and extent of white collar crime and crime against humanity. Powerlessness also corrupts (May, 1981). Without power, people often become alienated, relegating themselves to conditions of despair. Powerlessness may express itself in anger or apathy or some form of psychic depression. One sees this kind of corruption manifested among the poor, minorities, women, and the aged. Mass movements, non-cooperation, protests, and vigils are forms of resistance to overcome the corruption that comes from powerlessness. Through protests, people garner support for collective action that becomes the strategy for gaining access to knowledge so they can make meaningful demands upon those at the center of decision making.

integrating knowledge

The EA seminar is a way of accumulating and generating

knowledge for praxis--action and reflection. Praxis, like power, comes from integrating accumulated knowledge into our one's life experiences. The seminar not only gives one a basis for understanding macro-forces at work, but it provides one with foundations for praxis with reflection. The purpose of the seminar is not only to question the value judgments of the generators of knowledge, but to become aware of how certain value-latent knowledge leads to intellectual dishonesty. It questions basic assumptions that give such knowledge credibility.

With praxis may come action and reflection that break with the political conformity of the past. Too often, society is kept in a state of powerlessness because of implicit agreement with those who have power. People comply out of tradition, out of fear of sanction, or for the rewards to be gained from the establishment. Through knowledge of oppression, one may begin to take the first step for liberating oneself by refusing to cooperate with those responsible for oppressive conditions or those who thwart social change. Obviously, elite groups need cooperation of the people to rule and control natural resources and their distribution. Without such cooperation, they would have to abdicate their power over highly cherished resources, or at least they would find it more difficult to make decisions in support of their own narrowly defined interests. Through action and reflection, everyone can reclaim their power and create a new image of society as one of people who are loving, caring, smart, capable of understanding and making powerful decisions, and capable of following through on responsibilities. Despite overwhelming odds, some people--such as Gandhi, with his successful non-violent revolution against the British in India; Bertrand Russell, with his initiation of a meaningful test ban treaty; and Rosa Parks, with her refusal to vacate her bus seat for a white man in Montgomery, Alabama, thus sparking the civil rights movement--give testimony that individuals can make a difference. Change can start with one person staking claim to an issue and following through on responsible action. It can be done.

an organizational model

This seminar process is important in other ways, too. While the seminar is a model for teaching in university settings, it is also, by example, a model for organizational change and development, particularly in organizations with less than adequate resources. It can be used to build seminars around a given organizational problem in the absence of expensive consultants. Thus, it is important that students learn how to use this seminar skillfully, thus becoming less dependent upon outside resources. Moreover, former students can use each other as resources, sharing information across organizational lines.

Over the years, it has been found that students become more excited about education when they take charge of their own learning. With this comes a sense of empowerment that they can make a difference in the world in which they live.

CHAPTER 7

ADVOCATES
as futurists

Previous chapters have discussed and described EA not only as a critique of culture and political economy, but also as a means of developing skills to help act upon that critique. Often, one is rather explicit about the political, cultural, and economic dissatisfaction, but less clear about one's desires for the future. The world is becoming smaller due to technological advances in communication that extend awareness of the importance of bio-diversity and human cultures. EA has learned that it needs to find ways to minimize the impact of industrial development on the planet. The rate of biological extinction, the growth of poverty and despair, and the threat of nuclear war and global winters leads EA to believe that as advocates it is necessary to also become futurists.

Rapport (1986) has hailed culture as the greatest invention since life because it has allowed humans to adapt to any habitat on the face of the earth; yet, he questions whether culture has now become maladaptive. For the first time in human history, destructive forces such as global pollution and nuclear weapons have the potential to spell the end of homo sapiens, at which point, in the words of Shell (1982), "there will be no past, present, or future." Some biological determinists may argue that there has always been extinction and speciation, and that humans are no different from other species and subject to the same laws of nature. They may argue that one cannot stop the forces of evolution and that, in fact, something even more intelligent or healthier than humans may evolve ultimately out of human extinction and speciation. Although humans are no different from other species in many respects, we are different because we have culture; culture allows us to make choices about the future.

Cultural language allows us to transmit information across time and space and allows us to learn from past mistakes and knowledge. Furthermore, it allows us to escape not only from the here and now into the past, but from the here and now into the future. Through futuring, we can correct conditions in the present to prevent mistakes in the future, to be proactive rather than reactive. This means that we can control world events rather than allow world events to control us.

TREND ANALYSIS

Though language and the ability to project ourselves into the future, we can analyze societal trends. Some trends will lead to further depletion of world resources and the exploitation of people; other trends offer visions that are both refreshing and exciting. For the first time in a long time, there may be a trend toward peace on the globe; the world superpowers have relaxed tensions between themselves, and so have smaller, less-developed countries. Single-parent families, increased consciousness of the negative effects of environmental problems, comprehensive recycling, changing sexual preferences, vegetarianism, and holistic health care are all positive and perceptible trends of the future. Computers are beginning to redefine punctuality and education, the way we do business, and the way we receive mail and use the library. These uses of the computer can all be positive trends. Computers have many other potential uses--therapy sessions, dispute-resolution activities, group problem solving, and the enhancement of organizational communication.

a process of hoping

But a trend analysis is not enough. EA stresses that one needs to be clear about one's hopes for the future. Hopes for the future not only give us direction and meaning to our actions, but reaffirm our existence and express confidence in what we can become. Hope gives us a meaning for existence and a stake in the future. Hope is not only the nourishment that sustains us daily, but it affirms their existence (Skolimowski, no date). While hope is future-directed--we cannot hope for what is past, and there is no need to hope for what is present (Green, 1976)--it is an excellent guide to what is realistic and practical in the present. We may want things we can never have, but we cannot really hope beyond the boundaries of what is possible. So having hope is an important factor in planning for the future.

We are taught from our earliest years to learn from the past and from our mistakes, and much of EA's planning for the future is based on that principle. This is very useful when we attempt to choose among alternative courses of action. But often, the past constrains our ability to see new possibilities and visions of what could be. For example, imaging emphasizes creativity in making new choices rather than just selecting among the obvious and more traditional. Being goal oriented in planning may constrain creativity. This is not to say that creative solutions to problems cannot come from being goal directed, but goal directiveness may bind one too narrowly. Goals may limit the potential for creative solutions by preventing us from visualizing beyond the boundaries of predetermined goals. We may limits our potential by focusing too narrowly on goals that often have very specific parameters.

ENVIRONMENTAL ADVOCACY

new dimensions

We can learn from the future in much the same way we learn from the past, by adding new dimensions to such planning. Futuring is a planning process that allows us to focus on desired solutions rather than existing problems. It promotes a different approach to problem solving. It requires us to visualize the future as we would like it to be, without being constrained about how to achieve it. It is a freeing activity that has the potential to release creativity and visions that otherwise might never be.

Environmental advocates have been good at identifying "cutting" issues to empower people to make meaningful demands upon those at the center of decision-making power. We have been skillful in using issues not only to articulate problems that people can do something about, but also in winning victories to help build organizations of considerable power and strength. But such victories only push people into the near future--not too far from the present. For advocates to simple revisit the past for goal directedness or to "cut the issue" may bound our visions too narrowly. Yet, in most social change organizations, people are so inundated with issues, or direct action, or funding requirements, or organizational maintenance, that they seldom have time to do creative futuring. Given the state of the world, EA needs to find ways of extricating itself from the micro-level issues and begin to think globally. It not only needs to think globally and to act locally, but it needs to be able to image the future in such a way as to ensure human survival on this planet.

a planning technique

More and more, industry is using trend analysis and futuring as a planning technique to get a corner on the market for its products. Because events change so fast in modern society, industry needs to be well informed about trends and the future; its survival may depend upon it. Although few advocates and non-profit organizations have taken the time to future for the reasons mentioned previously, one such group has. The Environmental Action Group (EAG) was started 16 years ago, around the time of the energy crisis, by a Ph.D. graduate from the University of Michigan's School of Natural Resources. (While information in this chapter is based on real experiences, the name EAG is fictional, and the locations and other specifics about the group have been altered to conceal its actual identity.) While EAG's early mission was to slow energy development and ecological damage in the West by organizing and building coalitions between Native Americans and ranchers, it has shifted its focus over the years to providing technical assistance, training, and consultation to nonprofit organizations with missions of peace, environmental quality, and social justice.

FUTURING SESSIONS

EAG provides a number of services and training activities, such as proposal writing, decision making, long-range and inter-mediate-range planning, leadership development for nonprofits, budget and financial services, and conflict management. To carry out the mission of the organization, EAG requires funding from both foundations and the clients it serves. In order for it to be helpful to client organizations and itself, EAG wanted to use futuring as a planning tool. The EAG board and staff recently spent two days in a futuring exercise to help them plan. Although there were about 10 other people who participated in this session, most of the discussion in this chapter focuses on ideas from my involvement as an EAG board member and facilitator. Following is enough information to give the reader an idea of the session's content as well as its process, although the design of the futuring session is given here in modified form:

the futuring session

Minutes

15	Introduction
10	Brainstorm trends
35	Select five trends for analysis, using the trend analysis chart
15	Share results in total group
5	Introduction of hope
15	Hopes for the future to be written out
35	Share hopes in subgroups
15	Break (during break post hopes on wall for browsing)
5	Introduction: Image setting for the future
15	Images for the future (written out by individuals)
35	Share in small groups
30	Reports in total group (may want to look for common themes)
15	Write history of how you got to the future
35	Share results in sub groups
30	Share results in total group
15	Break
35	What are some strategies that come from our history?
	What are the ideologies behind our strategies?

Homework for Day II: A small group of two or three people take the mission statements and dissect them according to who does what, and how.

20	Report on committee work
35	Total group discussion: This is the time to object to what's in the chart or to make additions.

60	Participants in subgroups: A couple of groups could work on specific programs from the chart. Program goals can be established and a force-field analysis can be used to help strategize. Another group is to write a mission statement based upon the chart.
15	Break
60	Total group meeting for a progress report. Time for major and minor objections to the mission statement and work done in small groups.
35	Subgroups to continue to work on their projects
60	Total group meeting for a decision.

Following the introduction, participants in small groups brainstormed trends for the future and selected the five most important ones for trend analysis. They not only analyzed trends from social, demographic, technical, ecological, political, legal, and economic points of view, but they picked the worst- and the best-case scenarios that could impact EAG. A trend-analysis chart was given to participants to help them organize their thoughts. The trends identified by the group were such things as: single-parent families, peace in the world, environmental awareness, sexual preferences, rise of the radical right, cosmetic surgery, vegetarianism, holistic health care, civil rights, Japan as a first-rate world economic power, environmental degradation, street people, poverty, concentration of wealth, drug dependency, and others.

discussion

Following a group discussion on trends, a short presentation was made on hope and why it is important for futuring. Recognizing that the trends identified were not always positive, participants were asked to write down and discuss their hopes for the future. As a participant board observer, I wrote the following statement on hope; this hope statement and the imaging that follows do not necessarily reflect the ideas of EAG as a whole:

> Full employment is obsolete; full unemployment
> is in vogue. No matter what we have tried in the
> past, we have not been able to provide full
> employment for all our citizens. I hope for a
> guaranteed annual income available to 90
> percent of the people. We would, in turn, have
> to vote for 10 percent of the population to run
> government and industry. Of course, we will
> have to pay them more than the guaranteed
> income. I hope that we can make a paradigm
> shift where we build products for quality, more
> than for planned obsolescence. I hope that we
> can build mass-transit systems to last for 10,000
> years and houses to last the same length of
> time. Building products for quality and longevity
> will place less strain upon our natural resources.
> It will allow the guaranteed annual income to
> become more possible. There will be less work
> for people to do. However, my more immediate
> hopes are for workers to have a six-month paid
> sabbatical to use for personal growth and
> development.

Following a detailed discussion of the hopes of each board member, we then each wrote about our images of the future. The group projected itself to the year 2020 and described what was going on in the region which included EAG. Following is my personal image of the future:

> EAG has 20 staff members with an endowment
> of two million dollars. Staff are busy conferenc-
> ing on several computers simultaneously. I peer
> over the shoulder of Ellie, who is involved in a
> conference on peace. We now have a world

government where the United Nations has legitimate power to manage the globe for a fairer, more just and peaceful, and environmentally sound world. I move on to another computer where William is involved in a computer conference on the radical right. It seems they have been fairly active in the state of Idaho. Tony is actively involved in a conference on organizational change and development. The conference is being used to answer questions about conflict, decision making, planning, and futuring. Other conferences are going on, too. EAG has 20 computers and 4 laser writers.

I move on from the computers and observe in another room the putting together of video tapes on fundraising. This is a video studio with lots of equipment. I find out from one of the workers that this video tape will be duplicated and mailed to public-interest organizations in the region, to be followed up with a computer conference. Time spent on the road by field staff has been cut in half.

Suddenly, Ann appears. She has been on sabbatical for six months and will have to prepare herself to take charge of the computer conference on world peace, relieving Ellie for her sabbatical. I overhear a conversation in which Kate is talking with Daniel about her city health insurance policy. As I move through the organization, I also overhear Tom talking about putting money in the city-owned bank. He feels its a good thing to do. The bank makes money from its socially and environmentally sound investments to support city services.

Suddenly, my eye catches the headlines of the local newspaper. It states that the city-owned utility made a handsome profit. Another headline states that the city is banking its land to be leased to small businesses and used for affordable housing developments. This way the city can control both the direction and the rate of development.

As I begin to read more of the paper, I find that corporations are thriving, and there seem to be several worker-controlled industries. Most people are working six months out of the year. Then my consciousness is jarred when I hear a baby cry. My ears follow the sound, and I find that Mike Gibson's old office has been extended and turned into a day care center for EAG staff. There are a half dozen babies. I notice Asian, Native American, black, and white babies. I like what I see.

VISIONS EXPLAINED

Some participants found it rather challenging to commit their images of the future to paper. Often, it is fun to escape in the future and describe visions of an organization, community, or society that are different from the ones experienced today. A number of visions surfaced, and each was presented and discussed. Because my personal scenario of the future given here offers several images, some of them rather complex, each image deserves some more thorough discussion:

endowment

Two million dollar endowment: Historically it has been difficult for public-interest organizations to stabilize their funding over an extended period, particularly those nonprofits that are not membership based and are involved in technical assistance and training. Although computers have made it easier to write proposals, the fact remains that staff have to spend considerable time interpreting their program in face-to-face contact with funders or potential funders. Staff also have to spend time raising money--time that could be used to work with other organizations or for program training. Changing priorities of foundations and changing trends help to destabilize nonprofits, particularly those that are not membership based. A two million dollar endowment, perhaps over a period of 10 to 15 years, could stabilize funding for the EAG organization.

computer communication

Computer Conferencing: To service the Northwest requires that staff spend considerable time on the road because organizations are far apart. Time spent driving could be used more efficiently. Computer conferencing could allow staff to serve clients more efficiently and allow organizations to share resources. Although face-to-face contact with client groups is important and should be maintained, information can be exchanged and important training can take place using computer conferencing. Inexpensive, interactive computer conferencing can cut down travel time and augment service to nonprofits simultaneously. Computers have already changed the way work is done in organizations drastically, and will continue to do so.

world government

World Government: Historically, the globe has been managed in the interests of only a few developed countries (DCs). In the past, the spatial and temporal aspects of pollution were not extremely important because the by-products of technology only polluted a small area, taking the earth a relatively short period to heal itself. Today, however, the by-products of production, such as nuclear wastes (some of which have a half life of 100,000 years), global warming, acid rain, and the pollution of underground water reservoirs, have the potential to pollute and destroy large areas, taking the earth thousands of years to heal itself in some cases. Therefore, it is necessary to manage not just certain parts of the globe, but its entirety. Acid rain, global warming, nuclear wars, and radiation fall-out are not respectful of geographical boundaries. Poverty, despair, and oppression are becoming so widespread that there could be localized wars--wars that may trigger nuclear wars and nuclear winters the world over. The spatial and temporal aspects of global pollution may require the invoking of an international income tax to give some world organization, such as the United Nations, real power to manage the globe for environmental protection, economic betterment, and world peace. Is such an organization necessary? If so, how much power should it have? How should it be structured?

video training

Video Tapes: An inexpensive way of training people on a variety of issues is by mailing pre-recorded video tapes to nonprofits. Although the initial cost of video equipment may be expensive, making of video tapes for training and duplicating them will be relatively inexpensive, particularly if there is a video person on the staff. Because VCRs are so common, it would be economically feasible for client groups not only to own one, but to duplicate video tapes for their own library files before returning the original to EAG. Video tapes used in conjunction with computer conferencing will help EAG deliver quality services. This would be another way of saving time now spent traveling.

sabbaticals

Six-Month Sabbaticals: Full employment is becoming obsolete; full unemployment is in vogue. No matter what has been tried in the past, it has not been possible to provide full employment for all citizens. Society can be structured where 10 percent of the people are elected to run governmental affairs and industry, thus freeing the rest of us to pursue lifetime hobbies, education, and volunteer work. A guaranteed annual income will free people to travel around the nation, as well as the world over, to do interesting work of their choice. By changing production practices from planned obsolescence and conspicuous consumption to planned longevity and utilitarianism, from energy wastefulness to energy efficiency, society will place less strain on the biophysical environment and will use less resources. Mass transit systems, housing, computers, and household appliances can be built to last a hundred years or more. A six-month sabbatical is a move toward a new conception of work that defines full employment as both useful and ethical, that allows people opportunities to be creative and to experiment in a variety of ways. It would encourage people to engage in activities to enhance personal growth and development and to find new spiritual connections with mother earth.

public insurance

City Health Insurance Policy: Across the United States, there are thousands of people who are insurance poor. Due to the uneven development of the economy, the rich grow richer and the poor grow poorer, unable to provide a basic floor to protect themselves from unwanted illnesses. Why are there so many people in pain? Why do so many see their life savings succumb to catastrophic illnesses, some of which are connected to carcinogens found in the biophysical environment? No one in a society as wealthy as this should be without insurance protection; no one should risk losing their worldly goods or be dependent upon charity. A city-sponsored health insurance program is a must.

public banks

City-Owned Bank: Over years of neglect, many of the proud and largest cities have become wastelands of alienation and despair. Federal and state governments have failed to provide the necessary resources to revitalize cities, to make them fun, secure, and decent places to work and live. Rampant crime and delinquency--homicides, robberies, threats to physical well-being, and drug dependency--give modern cities a siege mentality; people are often scared and distrustful of one another. Corporations are closing their doors and moving to other parts of the nation and/ or to other countries for cheaper labor and resources, leaving thousands unemployed and an eroded tax base that strains already dwindling city services. One can no longer wait for federal or state governments to help restore these communities; the people must take charge and use tax dollars to form local city-owned banks to invest in socially and environmentally sound companies, with profits to be used to build vibrant, safe, and productive cities where people, regardless of their race, sex, origin, sexual preference, can realize their highest potential. The cities must be revitalized, or the siege mentality will continue to eat at the very core of our existence.

land banks

Land Banking: Although land banking is not a strategy often used in the United States country, it does have potential to help solve some of the problems that result from speculative growth in cities. Land speculation results in increased property values, which leads to the displacement of the poor, minorities, and senior citizens. The conception of the city as a growth machine catering to special interests must be changed. If a city landbank purchased key pieces of land to be leased to small business and individual home owners, it could control both the direction and the rate of development.

public utilities

City-Owned Utility: In many cities, considerable energy comes from outside the geopolitical boundaries. Although the price of energy has decreased since the 1973 oil embargo, unstable political situations or temporary or long-term shortages may increase the price of energy again. Energy conservation practices and a city-owned utility would result in big financial savings for citizens. With such a facility, the city could also use cogeneration--i.e., the use of energy developed for electricity to heat public buildings and homes, too. Cogeneration is used in many European cities, and it should be experimented with more in the U.S. than has been the case to date. Any profits from a city-owned utility would be deposited in the city-owned bank for future investments or for new and improved city services.

worker-controlled industry

Worker-Controlled Industries: While Americans are able to exercise democracy and vote for mayors and city council members who run the cities, they have not been able to extend democracy into the workplace to any significant extent. For eight working hours each day people have to give up their democracy and succumb to top-down, hierarchical, bureaucratic structures. Workers should not only be able to exercise democracy in the workplace, but they should work to create employment conditions that are safe and free from toxic and hazardous waste. There is little point in having economic democracy when the workplace contributes to chronic illness and a shortened life. It is not enough to have worker control without a deep respect for the environment, both in and out of the workplace.

child care

Day Care: Recognizing that children are society's most precious resource, it is necessary to provide a number of day care centers in and out of the work place. Quality day care will require trained staff with decent salaries and attractive working conditions. The importance of day care has been documented by research that indicates that disadvantaged children who have had the benefit of preschool day care are more likely to succeed in school. They also enjoy a better quality of life as young adults than those for whom day care is not available. Day care centers should be integrated by race and sex. Day care is an investment in the future of society. Day care services will allow adults in both dual-parent and single-parent families to realize their potential in the employment market. For each dollar spent on day care, seven dollars will be returned to society in such areas as reduced crime, reduced unemployment compensation, reduced welfare payments, and reduced need for remedial educational programs.

HISTORY
OF THE FUTURE

Such images of the future are indeed rich and provocative. These were my personal images, but others produced in the EAG sessions were just as interesting and varied considerably. Although the images written and spoken about in the sessions were not always complete, they did provide a good start for planning. The next phase of such training is to have people become "historians," writing the history of how they got to the year 2020. The history for my scenario of the future is outlined here:

History of How We Got to the Year 2020

1995   Wealth in this country has become even more skewed in favor of the rich. Street people are in abundance and homeless children are begging on the streets.

2000   More concern about growth and the depletion of the ozone layer, acid rain, poverty, and toxic waste.

2005   EAG begins to work on electoral campaigns and writes a document on eco-economic democracy. This document gives plans for day care, a landbank for the city of Boise, a community economic development corporation that focuses on local recycling industries, a city-owned bank, escrow accounts for affordable housing, a city-owned utility, energy conservation, city health insurance, and social and environmental impact statements. More people are concerned about public interest work. EAG receives its first million-dollar endowment from a progressive foundation.

2010   EAG has more staff. Three people are elected to the Boise City Council and endorse the eco-economic democracy document. On a world level, the United Nations has been given the power for global management. We have to pay a global income tax in order for the UN to respond to world poverty and environmental degradation and to work for peace. The UN is based on proportional representation by country.

2015   Both the Boise City Council and the state legislature have majorities that support the document.

2020   This brings us to the EAG scenario described above.

MISSION
STATEMENT

Writing a history of how one gets to the future can be quite informative. By writing a history, the group focuses on the time between the present and the future, on how the history will be written and what kinds of important events might affect the future, and on what is possible. The histories, which are read aloud to the group, often have implications for strategies for social change. One can learn by creating one's own history. From the history given here, one can glean strategies for change, as well as new projects or a mission statement for the organization. Or one can use information generated to discuss the political and theoretical framework that underlies images and strategies for the future. The following is an example of how to dissect a mission statement to help understand what is in it:

Mission Statement Chart

Who	Does What	For What Purpose	How
EAG	Conference on the Radical Right	To resolve racial hatred that causes tension and inequality among races	Computer conferences
	Conference on Organizations	To empower social and environmental organizations to carry out their mission	Computer conferences
	Conference on	To work for world	Computer

World Peace	peace so that resources for war can be used to improve the quality of life	conferences
Raise funds	To help carry out the mission of EAG	Computer, phone, video
Support City Health Insurance	To provide equal protection for quality medical care	Citizen tax for insurance
Support Landbank	To use land to control the rate and direction of growth	Tax write-off given to those who donate land to the city
Support City Bank	To increase city services to improve the quality of services	Establish a bank to be given to the city
Support City Utility	To provide moderate cost energy	Referendum
Support Day Care	To provide economic equality for men and women	Day care in the workplace
Support Worker Control of the Workplace	To provide pride and job security in the workplace	Worker buyouts

In an exercise like this one, each mission statement should be read and discussed. Then a group of two to three people should take all the mission statements and dissect them into a chart similar to the one given here. In the total group, the chart should be reviewed and participants should ask questions and check words or phrases about which they may feel uncomfortable. This is the time to add points that are not included on the chart. Once everyone feels comfortable with the chart, then the subgroup that created it, or perhaps a different one, takes the chart to write a mission statement for the organization. Other participants could form subgroups to discuss program ideas consistent with futuring. Both the mission statement and new program ideas can be worked on simultaneously. Periodically, subgroups should come together to check progress and to see if their projects or social change activities are consistent with the intent of the evolving mission statement.

the future one wants

As mentioned before, most people advocating social change find themselves in a reactive mode. Because the world is becoming dangerously close to ecological collapse and is being threatened by nuclear war, it is necessary to focus more on the future, to spend more time envisioning what kind of world one would like to live in--and then to work toward that end. EA seeks to find ways of extricating society from the shackles of the present so that everyone can devote the necessary time to creating and working for a future that will support people all over the world in sustainable environments. EA not only envisions what is possible and desirable, but it becomes a risk taker as well. Each EA group starts in its own community by thinking globally and acting locally. The goal is to make one's future what one wants it to be.

BIBLIOGRAPHY

Abercrombie, M. L. L. (1970). *Aims and Techniques of Group Teaching*. London: Society for Research into Higher Education, Ltd.

Alan, D. and Hanson, A., eds. (1972). *Recycle This Book!: Ecology, Society and Man*. Belmont, CA: Wadsworth Publishing Company.

Alderman, D. (1982). "Civil Right Activists Lead Anti-Dumping Protests in N.C." *The National Leader* (September 30).

Alinsky, S. (1969). *Reveille for Radicals*. New York: Vintage Books.

_____(1972). *Rules for Radicals: A Pragmatic Primer for Realistic Radicals*. New York: Vintage Books.

Allport, G. (1954). *The Nature of Prejudice*. Cambridge: Addison-Wesley.

Ash, R. (1972). *Social Movements in America*. Chicago: Markham Publishing Co.

Auvine, B. *et al*. (1977). *A Manual for Group Facilitators*. Madison, WI: The Center for Conflict Resolution.

Baran, P. and Sweezy, P. M. (1972). *Monopoly Capital: An Essay on the American Economic and Social Order*. New York: Modern Reader Paperbacks.

Barney, G. G. (1980). *The Global 2000 Report to the President of the U.S.: Entering the 21st Century*. New York: Pergamon Press.

Becker, H. (1967). "Whose Side Are We On?" *Social Problems*, Vol. 14.

Bello, W. *et al.* (1982). *Development Debacle: The World Bank in the Philippines*. San Francisco: Institute for Food and Development Policy.

Benello, C. G. (1971). "Group Organization and Social-Political Structure." In Benello, C. G. and Roussopoulos, D., eds. *The Case for Participatory Democracy*. New York: Viking.

Benne, K. (1959). "Some Ethical Problems in Group and Organizational Consultation." *Journal of Social Issues*, Vol. 15: 60-49.

Bennis, W. (1969). *Organizational Development: Its Nature, Origins, and Prospects*. Reading, MA: Addison-Wesley.

Berger, J. (1977). *Nuclear Power: The Unviable Option*. New York: Dell Publishing Co.

Bernstein, P. and Bowers, L. (1977). "Democratic Organization and Management." *Communities* (Summer).

Bluestone, B. and Bennett, H. (1982). *The Deindustrialization of America*. New York: Basic Books.

Bookchin, M. (1987). "Social Ecology Versus 'Deep Ecology,'" *Green Perspectives*, Numbers 4 and 5 (Summer).

———(1972). "Toward a Liberatory Technology." In Benello, C. G. and Roussopoulos, D., eds. *The Case for Participatory Democracy*. New York: The Viking Press.

———(1972). "The Ecological Use of Technology." In Benello, C. G. and Roussopoulos, D., eds. *The Case for a Participatory Democracy*. New York: The Viking Press.

———(1971). *Post Scarcity Anarchism*. Montreal: Black Rose Books.

Boulding, K. (1962). *Conflict and Defense: A General Theory*. New York: Harper Torchbooks.

Bowles, S. *et al.* (1983). *Beyond the Waste Land: A Democratic Alternative to Economic Decline*. New York: Anchor Press.

Bradford, L. (1976). *Making Meetings Work*. La Jolla, CA: University Associates.

———, Gibb, J., and Benne, K. (1964). *T-Group Theory and*

Laboratory Method. New York: Wiley.

Brager, G. and Specht, H. (1973). *Community Organizing*. New York: Columbia University Press.

Bryant, B. (1987). *Quality Circles: New Management Strategies for Schools*. Ann Arbor, MI: Caddo Gap Press.

_____ (1989). *Social Change, Energy, and Land Ethics*. Ann Arbor, MI: Caddo Gap Press.

_____ (1990). *Social and Environmental Change: A Manual for Community Organizing and Action*. Ann Arbor, MI: Caddo Gap Press.

Bullard, R. and Wright, B. (1986). "The Politics of Pollution: Implications for the Black Community." *Phylon*, Vol. 47.

Bullard, R. (1987). "Environmentalism and Politics of Equity: Emergent Trends in the Black Community." *Mid-America Review of Sociology*, Vol. 12.

Buttel, F. (1976). "Social Science and the Environment: Competing Theories." *Social Science Quarterly*, Vol. 57:2 (September).

_____ and Flinn, W. (1974). "The Structure of Support for the Environmental Movement, 1968-1970." *Rural Sociology*, Vol. 39:1 (Spring).

_____ and Frahm, A. (1982). "Appropriate Technology: Current Debate and Future Possibilities." *Humboldt Journal of Social Relations*, (Spring/Summer).

_____ and Larson, O. (1980). "Whither Environmentalism: The Future Political Path of the Environmental Movement." *Natural Resources Journal*, Vol. 20 (April).

Capra, F. and Spretnak, C. (1984). *Green Politics*. New York: E. P. Dutton, Inc.

Carnoy, M. and Shearer, D. (1980). *Economic Democracy: The Challenge of the 1980s*. New York: M. E. Sharpe, Inc.

Carr, W. and Kemmis, S. (1983). *Becoming Critical: Knowing Through Action Research*. Victoria, Australia: Deakin University Press.

Carroll, S. *et al.* (1972). *Revolution: A Quaker Prescription for a Sick Society*. Movement for a New Society Working Party

(July).

Carson, R. (1962). *Silent Spring*. Boston: Houghton Mifflin.

Catton, W. and Dunlap, R. E. (1980). "A New Ecological Paradigm for the Post-Exuberant Sociology." *American Behavioral Scientist*, Vol. 24:1 (February).

_____and Dunlap, R. E. (1978). "Environmental Sociology: A New Paradigm." *The American Sociologist*, Vol. 13 (February).

Cave, W. (1982). "Research Methods in the Social Sciences: Quantitative and Qualitative Modes Contrasted." Paper presented at the Finley Carpenter Research Conference, University of Michigan School of Education, Ann Arbor.

Center for Research on Learning and Teaching. (1978). "The Lecture." *Memo to the Faculty*, No. 60 (April). Ann Arbor, MI: University of Michigan Center for Research on Learning and Teaching.

Cloward, R. and Piven, F. (1982). *The New Class War: Reagan's Attack on the Welfare State and Its Consequences*. New York: Pantheon Books.

Cockcroft, J. D. et al., eds. (1972). *Dependence and Underdevelopment*. Garden City, New York: Doubleday Anchor.

The Commission for Racial Justice of the United Church of Christ. (1987). *Toxic Wastes and Race in the United States*.

Commoner, B. (1971). *The Closing Circle: Nature, Man and Technology*. New York: Bantam Books.

_____(1979). *The Politics of Energy*. New York: Alfred A. Knopf.

_____(1975). "How Poverty Breeds Over-Population." *Ramparts*, Vol. 13:10 (August-September).

_____(1976). *The Poverty of Power: Energy and the Economic Crisis*. New York: Bantam Books.

Coser, L (1956). *The Functions of Social Conflict*. New York: Bantam Books.

Crowfoot, J. and Bryant, B. (1980). "Environmental Advocacy: An Action Strategy for Dealing with Environmental Problems." *Journal of Environmental Education*, Vol. 11:3.

Cypher, J. M. (1981). "The Basic Economics of 'Rearming America.'" *Science for the People* (July/August).

Daly, H. (1973). *Toward a Steady-State Economy*. San Francisco: Freeman.

Dahrendorf, R. (1958). "Toward a Theory of Social Conflict." *Journal of Conflict Resolution*, No. 2:170-183.

Davidoff, P. *et al.* (1970). "Suburban Action: Advocate Planning for an Open Society." *AIP Journal* (January).

Devall, B. and Sessions, G. (1985). *Deep Ecology*. Salt Lake City: Gibbs M. Smith, Inc.

Dickerson, D. (1975). *The Politics of Alternative Technology*. New York: Universe.

Domhoff, W .G. (1979). *The Powers that Be: Processes of Ruling Class Domination in America*. New York: Vintage Books.

Dumas. L. (1979). "Economic Conversion: Cutting the Defense Budget Without Sacrificing Jobs." *Working Papers* (May-June).

Dunlap, R. E. (1980). "Paradigmatic Change in Social Science: From Human Exceptionalism to an Ecological Paradigm." *American Behavioral Scientist*, Vol. 214:1 (September/October).

Durkheim, E. (1964). *The Division of Labor in Society*. (Translated by Simpson, A.) New York: Free Press.

Ehrlich, P. R. (1971). *The Population Bomb*. New York: Ballantine Books.

England, R. and Bluestone, B. (1973). "Ecology and Class Conflict." In Daly, H., ed. *Toward a Steady-State Economy*. San Francisco: W. H. Freeman and Company.

_____and Bluestone, B. (1972). "Notes on the Political Economy of Pollution." In Allan, D. and Hanson, A., eds. *Recycle This Book!: Ecology, Society and Man*. Belmont, CA: Wadsworth Publishing Company.

Etzioni, A. (1964). *Modern Organizations*. San Francisco: Chandler.

Famelli, N. J. (1971). "Perilous Links Between Economic Growth, Justice and Ecology: A Challenge for Economic

Planners." *Environmental Affairs*, Vol. 1:2 (June).

Feller, G. *et al*, eds. (1981). *Peace and World Order Studies: A Curriculum Guide*. New York: Transnational Academic Program, Institute for World Order.

Frahm, A. and Buttel, F. H. (1982). "Appropriate Technology: Current Debate and Future Possibilities." *Humboldt Journal of Social Relations*, Vol. 9:2 (Spring/Summer).

Freidson, E. (1971). "Professionalism: The Doctor's Dilemma." *Social Policy*, (January/February).

Freire, P. (1974). *Pedagogy of the Oppressed*. New York: The Seabury Press.

Friends of the Earth. (1971) *Progress as if Survival Mattered*. San Francisco: Friends of the Earth.

_____ *et al*. (1982). *Ronald Reagan and the American Environment*. San Francisco: Friends of the Earth.

Gellon, M. (1970). "The Making of a Pollution-Industrial Complex." *Ramparts* (May).

George, S. (1977). *How the Other Half Dies: The Real Reasons for World Hunger*. Montclair, NJ: Alanheld, Osmun.

Goodman, R. (1979). *The Last Entrepreneurs: America's Regional Wars for Jobs and Dollars*. New York: Simon and Schuster.

Gouldner, A. (1968). "The Sociologist as Partisan: Sociology and the Welfare State." *American Sociologist*, Vol. 3.

_____(1962). "Anti-Minotaur: The Myth of a Value-Free Sociology." *Social Problems*, Vol. 9.

Gowan, S. *et al*. (1976). *Moving Toward a New Society*. Philadelphia: New Society Press.

Green, G. F. (1976). "Stories and Images of the Future." In Bundy, R., ed. *Images of the Future*. Buffalo, NY: Prometheus Books.

Grossman, R. and Daneker, G. (1977). *Jobs and Energy*. Washington, D.C.: Environmentalists for Full Employment.

Gyorgy, A. & Friends. (1979). *No Nukes: Everyone's Guide to Nuclear Power*. Boston: South End Press.

Hare, N. (1970). "Black Ecology." *The Black Scholar*, (April).

Hardin, G. (1973). "Tragedy of the Commons." In Daly, H., ed.

Toward a Steady-State Economy. San Francisco: W. H. Freeman and Company.

Haug, M. P. and Sussman, M. B. (1969). "Professional Autonomy and the Revolt of the Client." *Social Problems,* Vol. 17.

Hawley, C. (1979). "New Alchemy: A Guide to Cooperative Alternative Communities." *Journal of Cooperative Living* New Haven, CT: Community Publication Cooperative.

Hayes, D. (1976). "Energy: The Case for Energy Conservation." *World Watch Paper,* No. 4. World Watch Institute, (January).

Henderson, H. (1974). "The Entropy State." *Planning Review,* (April/May).

_____(1976). "The Coming Economic Transition." *Technological Forecasting and Social Change,* (March).

_____(1981). "New and Renewable Resource Technologies: Tools for a New International Economic Order." Unpublished paper presented at the United Nations' Conference on New and Renewable Sources of Energy, Nairobi, Kenya, (August 10-12).

Himes, J. (1966). "The Functions of Racial Conflict." *Social Forces,* 45:1.

Hodges, H. A. (1944). *Wilhelm Dilthey: An Introduction.* New York: Oxford University Press.

Horowitz, D., ed. (1969). *Corporations and the Cold War.* New York: Monthly Review Press.

Humphrey, C. R. and Buttel, F. R. (1982). *Environment, Energy, and Society.* Belmont, CA: Wadsworth Publishing Company.

Johnson, H. D. (1974). "Land Banking." *CoEvolution Quarterly,* (Summer).

Johnson, W. A. (1971). "The Guaranteed Income as an Environmental Measure." In Johnson and Hardesty, eds. *Economic Growth Vs. the Environment.* Belmont, CA: Wadsworth Publishing Company.

Jordan, V. (1978). "Energy Policy and Black People." *Vital Speeches of the Day.* Delivered at the Northern States Power Company Consumer/Utility Conference, Minneapolis, MN, (January 20).

Karenga, R. (1969). "The Black Community and the University: A Community Organizer's Perspective." In Robinson, *et al.*, eds. *Black Studies in the University: A Symposium*. New Haven, CT: Yale University Press.

Kelman, H. (1965). "Manipulation of Human Behavior: An Ethical Dilemma for the Social Scientists." *Journal of Social Issues*, Vol. 21.

Kieffer, C. H. (1981). *The Emergence of Empowerment: The Development of Participatory Competence Among Individuals in Citizen Organizations*. Vol. 1 and 2. (Dissertation). Ann Arbor: University Microfilm International.

Lappe, F. and Collins, J. (1977). *Food First: Beyond the Myth of Scarcity*. Boston: Houghton Mifflin.

_____(1971). *Diet for a Small Planet*. New York: Ballantine Books.

_____(1982). *Diet for a Small Planet--Tenth Anniversary Edition*. New York: Ballantine Books.

Leopold, D. (1974). *A Sand County Almanac*. New York: Sierra Club/Ballantine Book.

Lippitt, R. *et al.* (1958). *The Dynamics of Planned Change*. New York: Harcourt Brace Jovanovich.

Love, S. (1971). "Ecology and Social Justice: Is there a Conflict?" *Environmental Action*, (August 5).

Lovins, A. (1976). "Energy Strategy: The Road Not Taken?" *Foreign Affairs*, Vol. 55:1.

Mack, R. and Snyder, R. (1957). "The Analysis of Social Conflict--Toward an Overview and Synthesis." *Journal of Conflict Resolution*, No. 1, 212-248.

May, R. (1981). *Power and Innocence: A Search for the Sources of Violence*. New York: A Delta Book.

Meadows, D., *et al.* (1972). *The Limits to Growth*. New York: Universe Books.

Mermelstein, D., ed. (1973). *Economics, Mainstream Readings and Radical Critique*. New York: Random House.

Mesarovic, M. and Pestel, E. (1974). *Mankind at the Turning Point: The Second Report to the Club of Rome*. New York: E.

P. Dutton and Company, Inc./Readers Digest Press.

Miller, G. T., Jr. (1988). *Living in the Environment*, Fifth Edition. Belmont, CA: Wadsworth Publishing Company.

Mills, C. W. (1956). *The Power Elite*. New York: Oxford University Press.

Mitroff, I. V. (1974). *The Subjective Side of Science: A Philosophical Inquiry into the Psychology of the Apollo Moon Scientists*. New York: American Elsevier Publishing Company.

Montcrief, L. W. (1972). "The Cultural Basis of Our Environmental Crisis." In Dorfman, R. and Dorfman, N., eds. *Economics of the Environment*. New York: W. W. Norton.

Morrison, D. E., *et al.* (1972). "The Environmental Movement: Some Preliminary Observations and Predictions." In Burch, W., *et al.*, eds. *Social Behavior, Natural Resources and the Environment*. New York: Harper and Row.

_____(1980). "The Soft Cutting Edge of Environmentalism: Why and How the Appropriate Technology Notion Is Changing the Movement." *Natural Resources Journal*, Vol. 20 (April).

Napier, R. and Gershenfeld, M. K. (1973). *Groups: Theory and Experience*, Second Edition. Boston: Houghton Mifflin Company.

Newfield, J. (1966). *A Prophetic Minority*. New York: The New American Library.

O'Connor, J. (1973). *The Fiscal Crisis of the State*. New York: St. Martin's Press.

Olmsted, M. S. (1959). *The Small Group*. New York: Random House.

Ophuls, W. (1977). *Ecology and Politics of Scarcity: Prologue to a Political Theory of the Steady State*. San Francisco: W. H. Freeman.

Ouchi, W. (1981). *Theory Z*. New York: Avon.

Palmer, P. J. (1983). *To Know as We Are Known/A Spirituality of Education*. San Francisco: Harper and Row.

Pearl, A. (1970). "The More We Change, the Worse We Get. *Change*, (March/April).

Pearson, S. (1987). "Environmental Mediation and Empowerment." Unpublished term paper, University of Michigan School of Natural Resources, Ann Arbor, Michigan).

Peattie, L. R. (1968). "Reflections on Advocacy Planning." *AIP Journal*, (March).

_____(1970). "Drama and Advocacy Planning." *AIP Journal*, (November).

Penick, J., Jr. (1968). *Progressive Politics and Conservation*. Chicago: The University of Chicago Press.

Perelman, M. (1976). "Efficiency in Agriculture: The Economics of Energy." In Merrill, R., ed. *Radical Agriculture*. New York: Harper and Row.

Petulla, P. M. (1977). *American Environmental History*. San Francisco: Boyd and Fraser Publishing Company.

Pine, G. J. (1981). "Collaborative Action Research in School Counselling: The Integration of Research and Practice." *Personnel and Guidance Journal*, Vol. 59, pp. 495-501.

Piven, F. (1970). "Rejoinder: Disruption Is Still the Decisive Way." *Social Policy*, (July/August).

Rabb, E. and Lipset, S. (1971). "The Prejudiced Society." In Marx, G., ed. *Racial Conflict: Tension and Change in American Society*. Boston: Little Brown and Co.

Rankin, W. (1981). "Militarism, the Cities and the Poor." *The Nuclear Arms Race: Countdown to Disaster*. Cincinnati, OH: Forward Movement Publishers.

Rapport, R. (1986). "Lecture on Culture." Delivered at the School of Natural Resources, University of Michigan.

Reich, C. (1970). *The Greening of America*. New York: Random House.

Rifkin, J. (1981). *Entropy: A New World View*. New York: Bantam Books.

Robottom, I. M. (1983). "Familiarity and Contempt: Some Problems for Evaluation in Environmental Education." Unpublished paper prepared while serving as a visiting scholar at the University of Michigan School of Natural Resources, on leave from the School of Education, Deakin

University, Geelong 3217, Victoria, Australia.

Sale, K. (1987). "Ecofeminism--A New Perspective." *The Nation*, (September 26).

Satin, M. (1978). *New Age Politics: Healing Self and Society*. New York: A Delta Book.

Schein, E. H. (1965). *Organizational Psychology*. Englewood Cliffs, NJ: Prentice-Hall, Inc.

Schell, J. (1982). *The Fate of the Earth*. New York: Alfred A. Knopf.

Schnaiberg, A. (1980). *The Environment: From Surplus to Scarcity*. New York: Oxford University Press.

_____(1983). "Soft Energy and Hard Labor? Structural Restraints on the Transition to Appropriate Technology." In Summers, G. F., ed. *Technology and Rural Social Change*. Boulder, CO: Westview Press.

Schumacher, E. F. (1973). *Small Is Beautiful: Economics as if People Mattered*. New York: Harper and Row.

Schurmann, F. (1973). *Ideology and Organization in Communist China*. Berkeley: University of California Press.

Shaiken, H. (1980). "A Robot Is After Your Job." *The New York Times*, (September).

_____(1980). "Detroit Downsizes U.S. Jobs." *The Nation*, (October).

Sklar, H. (1980). *Trilateralism: The Trilateral Commission and Elite Planning for World Management*. Boston: South End Press.

Skolimowski, H. (no date). "Hope." Unpublished paper. Ann Arbor: University of Michigan.

Smith, J. K. (1983). "Quantitative Verses Qualitative Research: An Attempt to Clarify the Issue." *Educational Researcher*, Vol. 12:3 (March).

Smulyan, L. (1983). "Action Research on Change in Schools: A Collaborative Project." A paper presented at the Annual Meeting of the American Education Research Association, Montreal, Canada.

Stretton, H. (1976). *Capitalism, Socialism and the Environment*.

Cambridge, England: Cambridge University Press.

Tinbergen, J. (1976). *RIO: Reshaping the International Order.* New York: Dutton.

Toffler, A. (1980). *The Third Wave.* New York: William Morrow.

Tsongas, P. (1981). *The Road From Here: Liberalism and Realities in the 1980s.* New York: Alfred A. Knopf.

Vocations for Social Change. (no date). *No Bosses Here. A Manual on Working Collectively.* Cambridge, ME: Vocations for Social Change, 353 Broadway, Cambridge, ME 02139.

Weisberg, R. (1970). "The Politics of Ecology." *Liberation,* (January).

White, L. (1949). *The Science of Culture: The Study of Man and Civilization.* New York: American Book-Stratford Press.

Williams, R. M. (1958). "Value Orientation in American Society." In Stein, H. D. and Cloward R. A., eds. *Social Perspectives on Behavior.* New York: The Free Press of Glencoe, Inc.

Wolf, A. (1970). "The Perils of Professionalism." *Change,* (September-October).

Zwerdling, D. (1979). "Employee Ownership: How Well Is It Working?" *Working Papers,* (May-June).

OTHER BOOKS
by Bunyan Bryant
available from
Caddo Gap Press

Quality Circles: New Management Strategies for Schools.
1987. 85 pages, paperback. $8.95.

Social Change, Energy, and Land Ethics.
1989. 104 pages, paperback. $9.95.

**Social and Environmental Change: A Manual
for Community Organizing and Action.**
Second, Edition, 1990. 92 pages,
large format, paperback. $11.95.

To order or for more information contact:

Caddo Gap Press, Inc.
317 South Division Street, Suite 2
Ann Arbor, Michigan 48104

(313) 662-3604

ABOUT the author

Bunyan Bryant, a faculty member with the School of Natural Resources at the University of Michigan, teaches in the Program of Urban Technological and Environmental Planning. His courses include: "Small Group, Organization, and Advocacy Planning," "Social Change, Energy, and Land Ethics," and "Social Change and Natural Resources."

Bryant holds a Ph.D. degree in education and a M.S.W. degree in social work, both from the University of Michigan, and a B.S. degree from Eastern Michigan University, where he majored in social science with minors in biology and psychology. He has completed post-graduate study in Town and County Planning at the University of Manchester, England, and has extensive preparation in organizational training and development from the National Training Laboratory and Tavistock Laboratory.

While serving as a research project director at the Institute for Social Research at the University of Michigan in the late 1960s and early 1970s, **Bryant** designed and executed action research projects with school systems experiencing racial conflict. During his faculty service with the School of Natural Resources he has concentrated on research evaluation and training for public interest groups and nonprofit organizations.

Bryant feels that many educational, environmental, and ecological crises, such as the current difficulties experienced by family farms, are essentially socio-political and economic in character. In exploring such issues, he has served as a volunteer consultant and has encouraged his students to choose alternative careers--ones which will help people organize to improve their educational, environmental, and social conditions. **Bryant's** current research interests include race and the incidence of environmental hazards and comparing the effectiveness of various telecommunications modes for teaching, training, and organizing.

For information on the Environmental Advocacy field of academic study, contact:

Dr. **Bunyan Bryant**
School of Natural Resources
The University of Michigan
Ann Arbor, Michigan 48109-1115